Jonathan A. Barmak

Algebraic Topology
of Finite Topological Spaces
and Applications

 Springer

Jonathan A. Barmak
Departamento de Matemática
Fac. Cs. Exactas y Naturales
Universidad de Buenos Aires
Ciudad Universitaria Pabellón I
(1428) Ciudad de Buenos Aires
Argentina

ISBN 978-3-642-22002-9   e-ISBN 978-3-642-22003-6
DOI 10.1007/978-3-642-22003-6
Springer Heidelberg Dordrecht London New York

Lecture Notes in Mathematics ISSN print edition: 0075-8434
                             ISSN electronic edition: 1617-9692

Library of Congress Control Number: 2011934806

Mathematics Subject Classification (2010): 55-XX; 05-XX; 52-XX; 06-XX; 57-XX

Cover design: deblik, Berlin

Printed on acid-free paper

Springer is part of Springer Science+Business Media (www.springer.com)

A las dos bobes

Preface

There should be more math.
This could be mathier.

B.A. Summers

This book is a revised version of my PhD Thesis [5], supervised by Gabriel Minian and defended in March 2009 at the Mathematics Department of the Facultad de Ciencias Exactas y Naturales of the Universidad de Buenos Aires. Some small changes can be found here, following the suggestions of the referees and the editors of the LNM.

Gabriel proposed that we work together in the homotopy theory of finite spaces at the beginning of 2005, claiming that the topic had great potential and could be rich in applications. Very soon I became convinced of this as well. A series of notes by Peter May [51–53] and McCord and Stong's foundational papers [55, 76] were the starting point of our research. May's notes contain very interesting questions and open problems, which motivated the first part of our work.

This presentation of the theory of finite topological spaces includes the most fundamental ideas and results previous to our work and, mainly, our contributions over the last years. It is intended for topologists and combinatorialists, but since it is a self-contained exposition, it is also recommended for advanced undergraduate students and graduate students with a modest knowledge of Algebraic Topology.

The revisions of this book were made during a postdoc at Kungliga Tekniska högskolan, in Stockholm.

Acknowledgements

First and foremost, I would like to express my deepest gratitude to Gabriel Minian, the other person that played a leading role in the creation of this book, from its beginning as advisor of my undergraduate Thesis, until its completion with invaluable suggestions on the final presentation. I thank him for teaching me Algebraic Topology and for sharing with me his expertise. I gratefully acknowledge his encouragement, guidance and his confidence in

me. I thank him for developing with me the ideas presented in this book, and
for his unconditional support in each step of my career.

My PhD work at the Universidad de Buenos Aires was supported by a
fellowship of CONICET.

I want to thank Anders Björner and the combinatorics group at KTH,
where the revisions took place. My postdoc was supported by grant KAW
2005.0098 from the Knut and Alice Wallenberg Foundation.

I am grateful to the anonymous referees and to the editors of the LNM for
their suggestions that resulted in a better presentation of my work.

Thanks to my family and friends, especially to my parents and brother.
And Mor, and Bubu, of course.

Stockholm,   *Jonathan A. Barmak*
December 2010

Contents

Introduction

Most of the spaces studied in Algebraic Topology, such as CW-complexes or manifolds, are Hausdorff spaces. In contrast, finite topological spaces are rarely Hausdorff. A topological space with finitely many points, each of which is closed, must be discrete. In some sense, finite spaces are more natural than CW-complexes. They can be described and handled in a combinatorial way because of their strong relationship with finite partially ordered sets, but it is the interaction between their combinatorial and topological structures that makes them important mathematical objects. At first glance, one could think that such non Hausdorff spaces with just a finite number of points are uninteresting, but we will see that the theory of finite spaces can be used to investigate deep well-known problems in Topology, Algebra and Geometry.

In 1937, Alexandroff [1] showed that finite spaces and finite partially ordered sets (posets) are essentially the same objects considered from different points of view. However, it was not until 1966 that strong and deep results on the homotopy theory of finite spaces appeared, shaped in the two foundational and independent papers [76] and [55]. Stong [76] used the combinatorics of finite spaces to explain their homotopy types. This astounding article would have probably gone unnoticed if in the same year, McCord had not discovered the relationship between finite spaces and compact polyhedra. Given a finite topological space X, there exists an associated simplicial complex $\mathcal{K}(X)$ (the order complex) which has the same weak homotopy type as X, and, for each finite simplicial complex K, there is a finite space $\mathcal{X}(K)$ (the face poset) weak homotopy equivalent to K. Therefore, in contrast to what one could have expected at first sight, weak homotopy types of finite spaces coincide with homotopy types of finite CW-complexes. In this way, Stong and McCord put finite spaces in the game, showing implicitly that the interplay between their combinatorics and topology can be used to study homotopy invariants of well-known Hausdorff spaces.

Despite the importance of those papers, finite spaces remained in the shadows for many more years. During that time, the relationship between finite posets and finite simplicial complexes was exploited, but in most cases without knowing or neglecting the intrinsic topology of the posets. A clear example of this is the 1978 article of Quillen [70], who investigated the connection between algebraic properties of a finite group G and homotopy properties of the simplicial complex associated to the poset $S_p(G)$ of p-subgroups of G. In that beautiful article, Quillen left a challenging conjecture which remains open until this day. Quillen stated the conjecture in terms of the topology of the simplicial complex associated to $S_p(G)$. We will see that the finite space point of view adds a completely new dimension to his conjecture and allows one to attack the problem with new topological and combinatorial tools. We will show that the Whitehead Theorem does not hold for finite spaces: there are weak homotopy equivalent finite spaces with different homotopy types. This distinction between weak homotopy types and homotopy types is lost when we look into the associated polyhedra (because of the Whitehead Theorem) and, in fact, the essence of Quillen's conjecture lies precisely in the distinction between weak homotopy types and homotopy types of finite spaces.

In the last decades, a few interesting papers on finite spaces appeared [35, 65, 77], but the subject certainly did not receive the attention it required. In 2003, Peter May wrote a series of unpublished notes [51–53] in which he synthesized the most important ideas on finite spaces known until that time. In these articles, May also formulated some natural and interesting questions and conjectures which arose from his own research. May was one of the first to note that Stong's combinatorial point of view and the bridge constructed by McCord could be used together to attack problems in Algebraic Topology using finite spaces. My advisor, Gabriel Minian, chose May's notes, jointly with Stong's and McCord's papers, to be the starting point of our research on the Algebraic Topology of Finite Topological Spaces and Applications. This work, based on my PhD Dissertation defended at the Universidad de Buenos Aires in March 2009, is the first detailed exposition on the subject. In these notes I will try to set the basis of the theory of finite spaces, recalling the developments previous to ours, and I will exhibit the most important results of our work in the last years. The concepts and methods that we present in these notes are already being applied by many mathematicians to study problems in different areas.

Many results presented in this work are new and original. Various of them are part of my joint work with Gabriel Minian and appeared in our publications [7–11]. The results on finite spaces previous to ours appear in Chap. 1 and in a few other parts of the book where it is explicitly stated. Chapter 8, on equivariant simple homotopy types and Quillen's conjecture, and Chap. 11, on the Andrews-Curtis conjecture, contain some of the strongest results of these notes. These results are still unpublished.

New homotopical approaches to finite spaces that will not be treated in this book appeared for example in [22, 60] and more categorically oriented in [40, 73]. Applications of our methods and results to graph theory can be found in [19]. A relationship of finite spaces with toric varieties is discussed in [12].

In the first chapter we recall the correspondence between finite spaces and finite posets, the combinatorial description of homotopy types by Stong and the relationship between weak homotopy types of finite spaces and homotopy types of compact polyhedra found by McCord.

In Chap. 2 we give short basic proofs of many interesting original results. These include: the relationship between homotopy of maps between finite spaces and the discrete notion of homotopy for simplicial maps; an extension of Stong's ideas for pairs of finite spaces; the manifestation of finite homotopy types in the Hausdorff setting; a description of the fundamental group of a finite space; the realization of a finite group as the automorphism group of a finite space and classical constructions in the finite context, including a finite version of the mapping cylinder.

McCord found in [55] a *finite model* of the n-sphere S^n (i.e. a finite space weak homotopy equivalent to S^n) with only $2n + 2$ points. May conjectured in his notes that this space is, in our language, a *minimal finite model* of S^n, that is to say a finite model with minimum cardinality. In Chap. 3 we prove that May's conjecture is true. Moreover, the minimal finite model of S^n is unique up to homeomorphism (see Theorem 3.2.2). In this chapter we also study minimal finite models of finite graphs (CW-complexes of dimension 1) and give a full description of them in Theorem 3.3.7. In this case the uniqueness of the minimal finite models depends on the graph. The reason for studying finite models of spaces instead of finite spaces with the same homotopy type is that homotopy types of finite complexes rarely occur in the setting of finite spaces (see Corollary 2.3.4).

Given a finite space X, there exists a homotopy equivalent finite space X_0 which is T_0. That means that for any two points of X_0 there exists an open set which contains only one of them. Therefore, when studying homotopy types of finite spaces, we can restrict our attention to T_0-spaces.

In [76] Stong defined the notion of *linear* and *colinear points* of finite T_0-spaces, which we call *up beat* and *down beat points* following May's terminology. Stong proved that removing a beat point from a finite space does not affect its homotopy type. Moreover, two finite spaces are homotopy equivalent if and only if it is possible to obtain one from the other just by adding and removing beat points. On the other hand, McCord's results suggest that it is more important to study weak homotopy types of finite spaces than homotopy types. In this direction, we generalize Stong's definition of beat points introducing the notion of *weak point* (see Definition 4.2.2). If one removes a weak point x from a finite space X, the resulting space need not be homotopy equivalent to X, however we prove that in this case the inclusion $X \smallsetminus \{x\} \hookrightarrow X$ is a weak homotopy equivalence. As an application of this result, we exhibit an example (Example 4.2.1) of a finite space which is homotopically trivial, i.e. weak homotopy equivalent to a point, but which

is not contractible. This shows that the Whitehead Theorem does not hold
for finite spaces, not even for homotopically trivial spaces.

Osaki proved in [65] that if x is a beat point of a finite space X, there
is a simplicial collapse from the associated complex $\mathcal{K}(X)$ to $\mathcal{K}(X \smallsetminus \{x\})$.
In particular, if two finite spaces are homotopy equivalent, their associated
complexes have the same simple homotopy type. However, we noticed that
the converse is not true. There are easy examples of non-homotopy equivalent finite spaces with simple homotopy equivalent associated complexes.
The removing of beat points is a fundamental move in finite spaces, which
gives rise to homotopy types. We asked whether there exists another kind of
fundamental move in finite spaces, which corresponds exactly to the simple
homotopy types of complexes. We proved that the answer to this question
lies exactly in the notion of weak point. We say that there is a *collapse* from a
finite space X to a subspace Y if we can obtain Y from X by removing weak
points, and we say that two finite spaces have the same *simple homotopy type*
if we can obtain one from the other by adding and removing weak points.
We denote the first case with $X \searrow Y$ and the second case with $X \nearrow\!\!\!\searrow Y$.
The following result, which appears in Chap. 4, says that simple homotopy
types of finite spaces correspond precisely to simple homotopy types of the
associated complexes.

Theorem 4.2.11.

(a) *Let X and Y be finite T_0-spaces. Then, X and Y are simple homotopy
equivalent if and only if $\mathcal{K}(X)$ and $\mathcal{K}(Y)$ have the same simple homotopy
type. Moreover, if $X \searrow Y$ then $\mathcal{K}(X) \searrow \mathcal{K}(Y)$.*

(b) *Let K and L be finite simplicial complexes. Then, K and L are simple
homotopy equivalent if and only if $\mathcal{X}(K)$ and $\mathcal{X}(L)$ have the same simple
homotopy type. Moreover, if $K \searrow L$ then $\mathcal{X}(K) \searrow \mathcal{X}(L)$.*

This result allows one to use finite spaces to study problems of classical
simple homotopy theory. Indeed, we will use it to study the Andrews-Curtis
conjecture and we will use an equivariant version to investigate Quillen's
conjecture on the poset of p-subgroups of a finite group.

It is relatively easy to know whether two finite spaces are homotopy
equivalent using Stong's ideas, however it is very difficult (algorithmically
undecidable in fact) to distinguish if two finite spaces have the same weak
homotopy type. Note that this is as hard as recognizing if the associated
polyhedra have the same homotopy type. Our results on simple homotopy
types provide a first approach in this direction. If two finite spaces have
trivial Whitehead group, then they are weak homotopy equivalent if and
only if they are simple homotopy equivalent. In particular, a finite space
X is homotopically trivial if and only if it is possible to add and remove
weak points from X to obtain the singleton $*$. The importance of recognizing
homotopically trivial spaces will be evident when we study the conjecture of
Quillen. Note that the fundamental move in finite spaces induced by weak

points is easier to handle and describe than the simplicial one because it consists of removing just one single point of the space.

In the fourth section of Chap. 4 we study an analogue of Theorem 4.2.11 for simple homotopy equivalences. We give a description of the maps between finite spaces which correspond to simple homotopy equivalences at the level of complexes. The main result of this section is Theorem 4.4.12. In contrast to the classical situation where simple homotopy equivalences are particular cases of homotopy equivalences, homotopy equivalences between finite spaces are a special kind of simple homotopy equivalences.

As an interesting application of our methods on simple homotopy types, we will prove the following simple homotopy version of Quillen's famous Theorem A.

Theorem 4.5.2. *Let $\varphi : K \to L$ be a simplicial map between finite simplicial complexes. If $\varphi^{-1}(\sigma)$ is collapsible for every simplex σ of L, then $|\varphi|$ is a simple homotopy equivalence.*

In Chap. 5 we study the relationship between homotopy equivalent finite spaces and the associated complexes. The concept of contiguity classes of simplicial maps leads to the notion of *strong homotopy equivalence* (Definition 5.1.4) and *strong homotopy types* of simplicial complexes. This equivalence relation is generated by *strong collapses* which are more restrictive than the usual simplicial collapses. We prove the following result.

Theorem 5.2.1.

(a) If two finite T_0-spaces are homotopy equivalent, their associated complexes have the same strong homotopy type.

(b) If two finite complexes have the same strong homotopy type, the associated finite spaces are homotopy equivalent.

The notion of strong collapsibility is used to study the relationship between the contractibility of a finite space and that of its barycentric subdivision. This concept can be characterized using the nerve of a complex.

The fundamental moves described by beat or weak points are what we call *methods of reduction*. A reduction method is a technique that allows one to change a finite space to obtain a smaller one, preserving some topological properties, such as homotopy type, simple homotopy type, weak homotopy type or the homology groups. In [65], Osaki introduced two methods of this kind which preserve the weak homotopy type, and he asked whether these moves are effective in the following sense: given a finite space X, is it always possible to obtain a minimal finite model of X by applying repeatedly these methods? In Chap. 6 we give an example to show that the answer to this question is negative. In fact, it is a very difficult problem to find minimal finite models of spaces since this question is directly related to the problem of distinguishing weak homotopy equivalent spaces. Moreover, we prove that Osaki's methods of reduction preserve the simple homotopy type. In this

chapter we also study *one-point reduction methods* which consist of removing
just one point of the space. For instance, beat points and weak points lead
to one-point methods of reduction. In the second section of that chapter,
we define the notion of γ-*point* which generalizes the concept of weak point
and provides a more applicable method which preserves the weak homotopy
type. The importance of this new method is that it is almost the most general
possible one-point reduction method. More specifically, we prove the following
result.

Theorem 6.2.5. *Let X be a finite T_0-space, and $x \in X$ a point which is
neither maximal nor minimal and such that $X \smallsetminus \{x\} \hookrightarrow X$ is a weak homotopy
equivalence. Then x is a γ-point.*

In some sense, one-point methods are not sufficient to describe weak homotopy types of finite spaces. Concretely, if $x \in X$ is such that the inclusion
$X \smallsetminus \{x\} \hookrightarrow X$ is a weak homotopy equivalence, then $X \smallsetminus \{x\} \nearrow\hspace{-1.1em}\diagdown\, X$ (see
Theorem 6.2.8). Therefore, these methods cannot be used to obtain weak
homotopy equivalent spaces which are not simple homotopy equivalent.

Another of the problems originally stated by May in [52] consists in ex-
tending McCord's ideas in order to model, with finite spaces, not only
simplicial complexes, but general CW-complexes. We give an approach to this
question in Chap. 7, where the notion of *h-regular CW-complex* is defined. It
was already known that regular CW-complexes could be modeled by their
face posets. The class of h-regular complexes extends considerably the class of
regular complexes and we explicitly construct for each h-regular complex K, a
weak homotopy equivalence $K \to \mathcal{X}(K)$. Our results on h-regular complexes
allow the construction of new interesting examples of finite models. We also
apply these results to investigate quotients of finite spaces and derive a long
exact sequence of reduced homology for finite spaces.

Given a finite group G and a prime integer p, we denote by $S_p(G)$ the poset
of nontrivial p-subgroups of G. In [70], Quillen proved that if G has a nontrivial normal p-subgroup, then $\mathcal{K}(S_p(G))$ is contractible and he conjectured
the converse: if the complex $\mathcal{K}(S_p(G))$ is contractible, G has a nontrivial
p-subgroup. Quillen himself proved the conjecture for the case of solvable
groups, but the general problem still remains open. Some important advances
were achieved in [3]. As we said above, Quillen never considered $S_p(G)$ as a
topological space. In 1984, Stong [77] published a second article on finite
spaces. He proved some results on the equivariant homotopy theory of finite
spaces, which he used to attack Quillen's conjecture. He showed that G has a
nontrivial normal p-subgroup if and only if $S_p(G)$ is a contractible finite space.
Therefore, the conjecture can be restated in terms of finite spaces as follows:
$S_p(G)$ is contractible if and only if it is homotopically trivial. In Chap. 8 we
study an equivariant version of simple homotopy types of simplicial complexes
and finite spaces and we prove an analogue of Theorem 4.2.11 in this case.
Using this result we obtain some new formulations of the conjecture, which

are exclusively written in terms of simplicial complexes. One of these versions states that $\mathcal{K}(S_p(G))$ is contractible if and only if it has trivial equivariant simple homotopy type. We also obtain formulations of the conjecture in terms of the polyhedron associated to the much smaller poset $A_p(G)$ of the elementary abelian p-subgroups.

In Chap. 9 we describe homotopy properties of the so called *reduced lattices*, which are finite lattices with their top and bottom elements removed. We also introduce the $\mathcal{L}$ construction, which associates a new simplicial complex to a given finite space. This application, which is closed related to the nerve of a complex, was not included originally in my Dissertation [5]. We compare the homotopy type of a finite T_0-space X and the strong homotopy type of $\mathcal{L}(X)$. At the end of the chapter, another restatement of Quillen's conjecture is given using the complex $L_p(G) = \mathcal{L}(S_p(G))$. This version of the conjecture is closely related to the so called Evasiveness conjecture.

Chapter 10 is devoted to the study of fixed point sets of maps. We study the relationship between the fixed points of a simplicial automorphism and the fixed points of the associated map between finite spaces. We use this result to prove a stronger version of Lefschetz Theorem for simplicial automorphisms.

In the last chapter of these notes we exhibit some advances concerning the Andrews-Curtis conjecture. The geometric Andrews-Curtis conjecture states that if K is a contractible complex of dimension 2, then it 3-deforms to a point, i.e. it can be deformed into a point by a sequence of collapses and expansions which involve complexes of dimension not greater than 3. This long standing problem stated in the sixties, is closely related to Zeeman's conjecture and hence, to the famous Poincaré conjecture. With the proof of the Poincaré conjecture by Perelman, and by [30], we know now that the geometric Andrews-Curtis conjecture is true for *standard spines* (see [72]), but it still remains open for general 2-complexes. Inspired by our results on simple homotopy theory of finite spaces and simplicial complexes, we define the notion of *quasi constructible 2-complexes* which generalizes the concept of constructible complexes. Using techniques of finite spaces we prove that contractible quasi constructible 2-complexes 3-deform to a point. In this way we substantially enlarge the class of complexes which are known to satisfy the conjecture.

Throughout the book, basic results of Algebraic Topology will be assumed to be known by the reader. Nevertheless, we have included an appendix at the end of the notes, where we recall some basic concepts, ideas and classical results on simplicial complexes and CW-complexes that might be useful to the non-specialist.

Chapter 1
Preliminaries

In this chapter we will state some of the most important results on finite spaces which are previous to our work. These results can be summarized by the following three items:

1. The correspondence between finite topological spaces and finite partially ordered sets, first considered by Alexandroff in [1] in 1937.
2. The combinatorial description of homotopy types of finite spaces, discovered by Stong in his beautiful article [76] of 1966.
3. The connection between finite spaces and polyhedra, found by McCord [55] also in 1966.

A very nice exposition of the theory of finite spaces developed in the twentieth century can also be found in May's series of unpublished notes [51–53].

We will present Alexandroff and Stong's approaches to describe finite spaces, continuous maps, homotopies and connected components combinatorially. Then, we will compare weak homotopy types of finite spaces with homotopy types of compact polyhedra using McCord's results. Homotopy types of finite spaces were definitively characterized by Stong and homotopy equivalences between finite spaces are also well-understood. However, it is much more difficult to characterize weak homotopy equivalences. One of the most important tools to identify weak homotopy equivalences is the Theorem of McCord 1.4.2. However, we will see in following chapters that in some sense this result is not sufficient to describe all weak equivalences. The problem of distinguishing weak homotopy equivalences between finite spaces is directly related to the problem of recognizing homotopy equivalences between polyhedra.

J.A. Barmak, *Algebraic Topology of Finite Topological Spaces and Applications*, Lecture Notes in Mathematics 2032, DOI 10.1007/978-3-642-22003-6_1, © Springer-Verlag Berlin Heidelberg 2011

1.1   Finite Spaces and Posets

A *finite topological space* is a topological space with finitely many points and
a finite *preordered set* is a finite set with a transitive and reflexive relation.
We will see that finite spaces and finite preordered sets are basically the
same objects considered from different perspectives. Given a finite topological
space X, we define for every point $x \in X$ the *minimal open set* U_x as the
intersection of all the open sets which contain x. These sets are again open.
In fact arbitrary intersections of open sets in finite spaces are open. It is easy
to see that the minimal open sets constitute a basis for the topology of X.
Indeed, any open set U of X is the union of the sets U_x with $x \in U$. This
basis is called the *minimal basis* of X. Note that any other basis of X must
contain the minimal basis, since if U_x is a union of open sets, one of them
must contain x and then it coincides with U_x. We define a preorder on X by
$x \le y$ if $x \in U_y$.

If X is now a finite preordered set, one can define a topology on X given
by the basis $\{y \in X \mid y \le x\}_{x \in X}$. Note that if $y \le x$, then y is contained
in every basic set containing x, and therefore $y \in U_x$. Conversely, if $y \in U_x$,
then $y \in \{z \in X \mid z \le x\}$. Therefore $y \le x$ if and only if $y \in U_x$. This shows
that these two applications, relating topologies and preorders on a finite set,
are mutually inverse. This simple remark, made in first place by Alexandroff
[1], allows us to study finite spaces by combining Algebraic Topology with
the combinatorics arising from their intrinsic preorder structures.

The antisymmetry of a finite preorder corresponds exactly to the T_0 sep-
aration axiom. Recall that a topological space X is said to be T_0 if for any
pair of points in X there exists an open set containing one and only one of
them. Therefore finite T_0-spaces are in correspondence with finite partially
ordered sets (posets).

Example 1.1.1. Let $X = \{a, b, c, d\}$ be a finite space whose open sets are
$\emptyset$, $\{a, b, c, d\}$ $\{b, d\}$, $\{c\}$, $\{d\}$, $\{b, c, d\}$ and $\{c, d\}$. This space is T_0, and
therefore it is a poset. The first figure (Fig. 1.1) is a scheme of X with its
open sets represented by the interiors of the closed curves. A more useful
way to represent finite T_0-spaces is with their Hasse diagrams. The Hasse
diagram of a poset X is a digraph whose vertices are the points of X and
whose edges are the ordered pairs (x, y) such that $x < y$ and there exists
no $z \in X$ such that $x < z < y$. In the graphical representation of a Hasse
diagram we will not write an arrow from x to y, but a segment with y over
x (see Fig. 1.2).

If (x, y) is an edge of the Hasse diagram of a finite poset X, we say that y
covers x and write $x \prec y$.

An element x in a poset X is said to be *maximal* if $y \ge x$ implies $y = x$,
and it is a *maximum* if $y \le x$ for every $y \in X$. A finite poset has a maximum
if and only if there is a unique maximal element. The notions of *minimal
element* and *minimum* are dually defined. A *chain* in a poset is a subset of

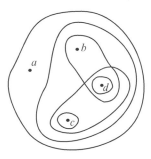

Fig. 1.1 Open sets of X

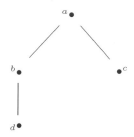

Fig. 1.2 Hasse diagram of X

elements which are pairwise comparable. An *antichain* is a subset of elements pairwise non-comparable.

Open sets of finite spaces correspond to *down-sets* and closed sets to *up-sets*. A subset U of a preordered set X is a down-set if for every $x \in U$ and $y \leq x$, it holds that $y \in U$. The notion of up-set is dually defined. If X is T_0, the open sets of X are in bijection with its antichains.

If x is a point of a finite space X, F_x denotes the closure of the set $\{x\}$ in X. Note that a point y is in F_x if and only if $x \in U_y$. Therefore, $F_x = \{y \in X \mid y \geq x\}$. If a point x belongs to finite spaces X and Y, we write U_x^X, U_x^Y, F_x^X and F_x^Y so as to distinguish whether the minimal open sets and closures are considered in X or in Y.

Note that the set of closed subspaces of a finite space X is also a topology on the underlying set of X. The finite space with this topology is the *opposite* of X (or *dual*) and it is denoted by X^{op}. The order of X^{op} is the inverse order of X. If $x \in X$, then $U_x^{X^{op}} = F_x^X$.

If A is a subspace of a topological space X, the open sets of A are the intersections of open sets of X with A. In particular, if X is finite and $a \in A$, $U_a^A = U_a^X \cap A$. If X and Y are two topological spaces, a basis for the product topology in the cartesian product $X \times Y$ is given by products $U \times V$ with U an open subset of X and V open in Y. Then, if X and Y are finite and $(x,y) \in X \times Y$, $U_{(x,y)} = U_x \times U_y$. Therefore, we have the following

Remark 1.1.2.

(a) Let A be a subspace of a finite space X and let $a, a' \in A$. Then $a \leq_A a'$ if and only if $a \leq_X a'$. Here $\leq_A$ denotes the preorder corresponding to the subspace topology of A and $\leq_X$ the corresponding to the topology of X.

(b) Let X and Y be two finite spaces and let $(x, y), (x', y')$ be two points in the cartesian product $X \times Y$ considered with the product topology. Then $(x, y) \leq (x', y')$ if and only if $x \leq x'$ and $y \leq y'$.

1.2   Maps, Homotopies and Connectedness

We will see that the notions of morphisms in finite spaces and in finite preordered sets are exactly the same. Moreover, the connected components of a finite space coincide with its path connected components and with the connected components of the corresponding finite preordered set. This result will be used to give a combinatorial description of homotopies.

A function $f : X \to Y$ between two preordered sets is *order preserving* if $x \leq x'$ implies $f(x) \leq f(x')$ for every $x, x' \in X$.

Proposition 1.2.1. *A function $f : X \to Y$ between finite spaces is continuous if and only if it is order preserving.*

Proof. Suppose f is continuous and $x \leq x'$ in X. Then $f^{-1}(U_{f(x')}) \subseteq X$ is open and since $x' \in f^{-1}(U_{f(x')})$, $x \in U_{x'} \subseteq f^{-1}(U_{f(x')})$. Therefore $f(x) \leq f(x')$.

Now assume that f is order preserving. To prove that f is continuous it suffices to show that $f^{-1}(U_y)$ is open for every set U_y of the minimal basis of Y. Let $x \in f^{-1}(U_y)$ and let $x' \leq x$. Then $f(x') \leq f(x) \leq y$ and $x' \in f^{-1}(U_y)$. This proves that $f^{-1}(U_y)$ is a down-set.   □

If $f : X \to Y$ is a function between finite spaces, the map $f^{op} : X^{op} \to Y^{op}$ is the map which coincides with f in the underlying sets. It easy to see that f is continuous if and only if f^{op} is continuous.

Remark 1.2.2. If X is a finite space, a one-to-one continuous map $f : X \to X$ is a homeomorphism. In fact, since f is a permutation of the set X, there exists $n \in \mathbb{N} = \mathbb{Z}_{\geq 1}$ such that $f^n = 1_X$.

Lemma 1.2.3. *Let x, y be two comparable points of a finite space X. Then there exists a path from x to y in X, i.e. a map α from the unit interval I to X such that $\alpha(0) = x$ and $\alpha(1) = y$.*

Proof. Assume $x \leq y$ and define $\alpha : I \to X$, $\alpha(t) = x$ if $0 \leq t < 1$, $\alpha(1) = y$. If $U \subseteq X$ is open and contains y, then it contains x also. Therefore $\alpha^{-1}(U)$ is one of the following sets, $\emptyset$, I or $[0, 1)$, which are all open in I. Thus, α is a path from x to y.   □

Let X be a finite preordered set. A *fence* in X is a sequence $x_0, x_1, \ldots, x_n$ of points such that any two consecutive are comparable. X is *order-connected* if for any two points $x, y \in X$ there exists a fence starting in x and ending in y.

Proposition 1.2.4. *Let X be a finite space. Then the following are equivalent:*

1. *X is a connected topological space.*
2. *X is an order-connected preorder.*
3. *X is a path-connected topological space.*

Proof. If X is order-connected, it is path-connected by Lemma 1.2.3. We only have to prove that connectedness implies order-connectedness. Suppose X is connected and let $x \in X$. Let $A = \{y \in X \mid \text{there is a fence from } x \text{ to } y\}$. If $y \in A$ and $z \le y$, then $z \in A$. Therefore A is a down-set. Analogously, it is an up-set and then, $A = X$.   □

Recall that if X and Y are two topological spaces, the *compact-open* topology in the set Y^X of maps from X to Y is the topology whose subbase is given by the sets $S(K, W) = \{f \in Y^X \mid f(K) \subseteq W\}$ where K is a compact subset of X and W is an open subset of Y.

If X and Y are finite spaces we can consider the finite set Y^X of continuous maps from X to Y with the pointwise order: $f \le g$ if $f(x) \le g(x)$ for every $x \in X$.

Proposition 1.2.5. *Let X and Y be two finite spaces. Then pointwise order on Y^X corresponds to the compact-open topology.*

Proof. Let $S(K, W) = \{f \in Y^X \mid f(K) \subseteq W\}$ be a set of the subbase of the compact-open topology, where K is a (compact) subset of X and W an open set of Y. If $g \le f$ and $f \in S(K, W)$, then $g(x) \le f(x) \in W$ for every $x \in K$ and therefore, $g \in S(K, W)$. Thus, $S(K, W)$ is a down-set in Y^X. Conversely, if $f \in Y^X$, $\{g \in Y^X \mid g \le f\} = \bigcap_{x \in X} S(\{x\}, U_{f(x)})$. Therefore both topologies coincide.   □

The exponential law for sets claims that if X, Y and Z are three sets, there is a natural bijection ϕ between the set of functions $f : X \times Z \to Y$ and the set of functions $Z \to Y^X$ which is given by $\phi(f)(z)(x) = f(x, z)$. If X, Y and Z are now topological spaces and Y^X is given the compact-open topology, the continuity of a function $f : X \times Z \to Y$ implies the continuity of $\phi(f) : Z \to Y^X$. However, the converse is not true in general. If X is a locally compact Hausdorff space, ϕ does define a bijection between the set of continuous maps $X \times Z \to Y$ and the set of continuous maps $Z \to Y^X$. More generally, it is enough that every point x of X has a basis of compact neighborhoods, or in other words, that for every open set U containing x there exists a compact neighborhood of x contained in U (see [27] or [62, Theorems 46.10 and 46.11]). If X is a finite space, every subspace of X is compact and this condition is trivially satisfied. In particular, if X is a finite space and Y is any topological space, there is a natural correspondence between the set of homotopies $\{H : X \times I \to Y\}$ and the set of paths $\{\alpha : I \to Y^X\}$. From now on we consider the map spaces Y^X with the compact-open topology, unless we say otherwise.

Given two maps $f, g : X \to Y$ between topological spaces, we will write $f \simeq g$ if they are homotopic. Moreover, if they are homotopic relative to a subspace $A \subseteq X$, we will write $f \simeq g$ rel A.

Corollary 1.2.6. *Let $f, g : X \to Y$ be two maps between finite spaces. Then $f \simeq g$ if and only if there is a fence $f = f_0 \le f_1 \ge f_2 \le \ldots f_n = g$. Moreover, if $A \subseteq X$, then $f \simeq g$ rel A if and only if there exists a fence $f = f_0 \le f_1 \ge f_2 \le \ldots f_n = g$ such that $f_i|_A = f|_A$ for every $0 \le i \le n$.*

Proof. There exists a homotopy $H : f \simeq g$ rel A if and only if there is a path $\alpha : I \to Y^X$ from f to g such that $\alpha(t)|_A = f|_A$ for every $0 \le t \le 1$. This is equivalent to saying that there is a path $\alpha : I \to M$ from f to g where M is the subspace of Y^X of maps which coincide with f in A. By Proposition 1.2.4 this means that there is a fence from f to g in M. The order of M is the one induced by Y^X, which is the pointwise order by Proposition 1.2.5. $\qquad\square$

Corollary 1.2.7. *Let $f, g : X \to Y$ be two maps between finite spaces. Then $f \simeq g$ if and only if $f^{op} \simeq g^{op}$. In particular, f is a homotopy equivalence if and only if $f^{op} : X^{op} \to Y^{op}$ is a homotopy equivalence and two finite spaces are homotopy equivalent if and only if their duals are homotopy equivalent.*

Remark 1.2.8. Any finite space X with maximum or minimum is contractible since, in that case, the identity map 1_X is comparable with a constant map c and therefore $1_X \simeq c$.

For example, the space of Fig. 1.2 has a maximum and therefore it is contractible.

Note that if X and Y are finite spaces and Y is T_0, then Y^X is T_0 since $f \le g$, $g \le f$ implies $f(x) = g(x)$ for every $x \in X$.

1.3 Homotopy Types

In this section we will recall the beautiful ideas of Stong [76] about homotopy types of finite spaces. Stong introduced the notion of *linear* and *colinear points*, which we will call *up beat* and *down beat points* following May's nomenclature [51]. Removing such points from a finite space does not affect its homotopy type. Therefore any finite space is homotopy equivalent to a space without beat points, which is called a *minimal finite space*. The Classification Theorem will follow from this remarkable result: two minimal finite spaces are homotopy equivalent only if they are homeomorphic.

The next result essentially shows that, when studying homotopy types of finite spaces, we can restrict ourselves to T_0-spaces. Recall that if $\sim$ is a relation on a topological space X, then the quotient topology on $X/\sim$ is the final topology with respect to the quotient map $q : X \to X/\sim$. In other words, $U \subseteq X/\sim$ is open if and only if $q^{-1}(U)$ is open in X. The quotient topology satisfies the following property: a function f from $X/\sim$ to any topological space Y is continuous if and only if the composition $fq : X \to Y$ is continuous.

Proposition 1.3.1. *Let X be a finite space. Let X_0 be the quotient $X/\sim$ where $x \sim y$ if $x \leq y$ and $y \leq x$. Then X_0 is T_0 and the quotient map $q : X \to X_0$ is a homotopy equivalence.*

Proof. Take any section $i : X_0 \to X$, i.e. $qi = 1_{X_0}$. The composition iq is order preserving and therefore i is continuous. Moreover, since $iq \leq 1_X$, i is a homotopy inverse of q.

Let $x, y \in X_0$ such that $q(x) \leq q(y)$, then $x \leq iq(x) \leq iq(y) \leq y$. If in addition $q(y) \leq q(x)$, $y \leq x$ and then $q(x) = q(y)$. Therefore the preorder of X_0 is antisymmetric. □

Remark 1.3.2. Note that the map $i : X_0 \to X$ of the previous proof is a subspace map since $qi = 1_{X_0}$. Moreover, since $iq \leq 1_X$ and the maps iq and 1_X coincide on X_0, then by Corollary 1.2.6, $iq \simeq 1_X$ rel X_0. Therefore X_0 is a strong deformation retract of X.

Definition 1.3.3. A point x in a finite T_0-space X is a *down beat point* if x covers one and only one element of X. This is equivalent to saying that the set $\hat{U}_x = U_x \smallsetminus \{x\}$ has a maximum. Dually, $x \in X$ is an *up beat point* if x is covered by a unique element or equivalently if $\hat{F}_x = F_x \smallsetminus \{x\}$ has a minimum. In any of these cases we say that x is a *beat point* of X.

It is easy to recognize beat points looking into the Hasse diagram of the space. A point $x \in X$ is a down beat point if and only if its indegree in the digraph is 1 (there is one and just one edge with x at its top). It is an up beat point if and only if its outdegree is 1 (there is one and only one edge with x at the bottom). In the example of Fig. 1.2 in page 3, a is not a beat point: it is not a down beat point because there are two segments with a at the top and it is not an up beat point either because there is no segment with a at the bottom. The point b is both a down and an up beat point, and c is an up beat point but not a down beat point.

If X is a finite T_0-space, and $x \in X$, then x is a down beat point of X if and only if it is an up beat point of X^{op}. In particular x is a beat point of X if and only if it is a beat point of X^{op}.

Proposition 1.3.4. *Let X be a finite T_0-space and let $x \in X$ be a beat point. Then $X \smallsetminus \{x\}$ is a strong deformation retract of X.*

Proof. Assume that x is a down beat point and let y be the maximum of $\hat{U}_x$. Define the retraction $r : X \to X \smallsetminus \{x\}$ by $r(x) = y$. Clearly, r is order-preserving. Moreover if $i : X \smallsetminus \{x\} \hookrightarrow X$ denotes the canonical inclusion, $ir \leq 1_X$. By Corollary 1.2.6, $ir \simeq 1_X$ rel $X \smallsetminus \{x\}$. If x is an up beat point the proof is similar.   □

Definition 1.3.5. A finite T_0-space is a *minimal finite space* if it has no beat points. A *core* of a finite space X is a strong deformation retract which is a minimal finite space.

By Remark 1.3.2 and Proposition 1.3.4 we deduce that every finite space has a core. Given a finite space X, one can find a T_0-strong deformation retract $X_0 \subseteq X$ and then remove beat points one by one to obtain a minimal finite space. The notable property about this construction is that in fact the core of a finite space is unique up to homeomorphism, moreover: two finite spaces are homotopy equivalent if and only if their cores are homeomorphic.

Theorem 1.3.6. *Let X be a minimal finite space. A map $f : X \to X$ is homotopic to the identity if and only if $f = 1_X$.*

Proof. By Corollary 1.2.6 we may suppose that $f \leq 1_X$ or $f \geq 1_X$. Assume $f \leq 1_X$. Let $x \in X$ and suppose by induction that $f|_{\hat{U}_x} = 1_{\hat{U}_x}$. If $f(x) \neq x$, then $f(x) \in \hat{U}_x$ and for every $y < x$, $y = f(y) \leq f(x)$. Therefore, $f(x)$ is the maximum of $\hat{U}_x$ which is a contradiction since X has no down beat points. Then $f(x) = x$. The case $f \geq 1_X$ is similar.   □

Corollary 1.3.7 (Classification Theorem). *A homotopy equivalence between minimal finite spaces is a homeomorphism. In particular the core of a finite space is unique up to homeomorphism and two finite spaces are homotopy equivalent if and only if they have homeomorphic cores.*

Proof. Let $f : X \to Y$ be a homotopy equivalence between finite spaces and let $g : Y \to X$ be a homotopy inverse. Then $gf = 1_X$ and $fg = 1_Y$ by Theorem 1.3.6. Thus, f is a homeomorphism. If X_0 and X_1 are two cores of a finite space X, then they are homotopy equivalent minimal finite spaces, and therefore, homeomorphic. Two finite spaces X and Y have the same homotopy type if and only if their cores are homotopy equivalent, but this is the case only if they are homeomorphic.   □

Example 1.3.8. Let X and Y be the following finite T_0-spaces:

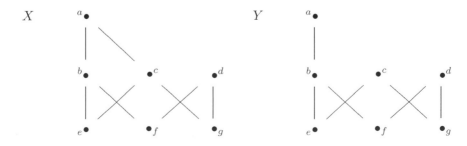

The following sequence of figures, shows how to obtain the core of X removing beat points. Note that b is an up beat point of X, c is an up beat point of $X \smallsetminus \{b\}$ and e an up beat point of $X \smallsetminus \{b, c\}$. The subspace $X \smallsetminus \{b, c, e\}$ obtained in this way is a minimal finite space and then it is the core of X.

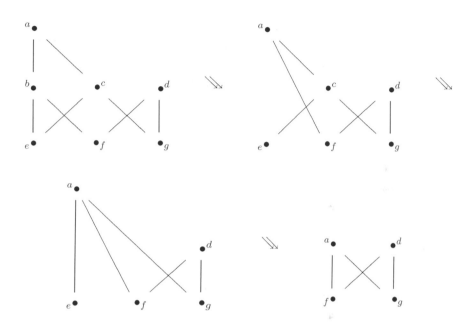

On the other hand, a is a beat point of Y and $Y \smallsetminus \{a\}$ is minimal. Therefore the cores of X and Y are not homeomorphic, so X and Y are not homotopy equivalent.

By the Classification Theorem, a finite space is contractible if and only if its core is a point. Therefore any contractible finite space has a point which is a strong deformation retract. This property is false in general for non-finite spaces (see [38, Exercise 6(b), p. 18]). It is not true either that every point in a contractible finite space X is a strong deformation retract of X (see Example 2.2.6).

Note that the core X_c of a finite space X is the smallest space homotopy equivalent to X. If Y is another finite space homotopy equivalent to X, then the core of Y must be homeomorphic to X_c and it has at most as many points as Y.

In [76] Stong also gives a homeomorphism classification result considering matrix representations of finite spaces. A similar approach can be found also in the survey [43].

To finish this section, we exhibit the following characterization of minimal finite spaces which appears in [11].

Proposition 1.3.9. *Let X be a finite T_0-space. Then X is a minimal finite space if and only if for all $x, y \in X$, if every $z \in X$ comparable with x is comparable with y, then $x = y$.*

Proof. If X is not minimal, there exists a beat point x. Without loss of generality assume that x is a down beat point. Let y be the maximum of $\hat{U}_x$. Then if $z \geq x$, $z \geq y$ and if $z < x$, $z \leq y$.

Conversely, suppose that there exist $x \neq y$ such that every element comparable with x is also comparable with y. In particular x is comparable with y. We may assume that $x > y$. Let $A = \{z \in X \mid z > y$ and for every $w \in X$ comparable with z, w is comparable with $y\}$. This set is nonempty since $x \in A$. Let x' be a minimal element of A. We show that x' is a down beat point with $y = \max(\hat{U}_{x'})$. Let $z < x'$, then z is comparable with y since $x' \in A$. Suppose $z > y$. Let $w \in X$. If $w \leq z$, then $w \leq x'$ and so, w is comparable with y. If $w \geq z$, $w \geq y$. Therefore $z \in A$, contradicting the minimality of x'. Then $z \leq y$. Therefore y is the maximum of $\hat{U}_{x'}$. $\square$

1.4   Weak Homotopy Types: The Theory of McCord

In the previous section we have studied homotopy types of finite spaces. On the other hand we will see in the next chapter, that Hausdorff spaces do not have in general the homotopy type of any finite space. However finite CW-complexes do have the same *weak homotopy types* as finite spaces. In 1966 McCord proved that every compact polyhedron has an associated finite space with the same weak homotopy type and every finite space has a weak equivalent associated polyhedron.

Recall that a continuous map $f : X \to Y$ between topological spaces is said to be a weak homotopy equivalence if it induces isomorphisms in all homotopy groups, i.e. if $f_* : \pi_0(X) \to \pi_0(Y)$ is a bijection and the maps

$$f_* : \pi_n(X, x_0) \to \pi_n(Y, f(x_0))$$

are isomorphisms for every $n \geq 1$ and every base point $x_0 \in X$. Homotopy equivalences are weak homotopy equivalences, and the Whitehead Theorem [38, Theorem 4.5] claims that any weak homotopy equivalence between CW-complexes is a homotopy equivalence. However, in general, these two concepts differ. We will show examples of weak homotopy equivalences between two finite spaces and between a polyhedron and a finite space which are not homotopy equivalences.

Weak homotopy equivalences satisfy the so called 2-out-of-3 property. That means that if f and g are two composable maps and 2 of the 3 maps f, g, gf are weak homotopy equivalences, then so is the third. Moreover if f and g are two homotopic maps and one is a weak homotopy equivalence, then so is the other. Any weak homotopy equivalence $f : X \to Y$ between topological spaces induces isomorphisms $f_* : H_n(X; G) \to H_n(Y; G)$ between the homology groups, for every $n \geq 0$ and every coefficient group G [38, Proposition 4.21]. The singular homology groups with integer coefficients of a space X will be denoted as usual by $H_n(X)$.

The Theorem of McCord 1.4.2 plays an essential role in the homotopy theory of finite spaces. This result basically says that if a continuous map is locally a weak homotopy equivalence, then it is a weak homotopy equivalence itself. The original proof by McCord can be found in [55, Theorem 6] and it is based on an analogous result for quasifibrations by Dold and Thom. An alternative proof for finite covers can also be obtained from [38, Corollary 4K.2].

Definition 1.4.1. Let X be a topological space. An open cover $\mathcal{U}$ of X is called a *basis like open cover* if $\mathcal{U}$ is a basis for a topology in the underlying set of X (perhaps different from the original topology). This is equivalent to saying that for any $U_1, U_2 \in \mathcal{U}$ and $x \in U_1 \cap U_2$, there exists $U_3 \in \mathcal{U}$ such that $x \in U_3 \subseteq U_1 \cap U_2$.

For instance, if X is a finite space, the minimal basis $\{U_x\}_{x \in X}$ is a basis like open cover of X.

Theorem 1.4.2 (McCord). *Let X and Y be topological spaces and let $f : X \to Y$ be a continuous map. Suppose that there exists a basis like open cover $\mathcal{U}$ of Y such that each restriction*

$$f|_{f^{-1}(U)} : f^{-1}(U) \to U$$

is a weak homotopy equivalence for every $U \in \mathcal{U}$. Then $f : X \to Y$ is a weak homotopy equivalence.

Example 1.4.3. Consider the following map between finite spaces

defined by $f(a_1) = f(a_2) = f(a_3) = a$, $f(b) = b$, $f(c) = c$, $f(d) = d$. It is order preserving and therefore continuous. Moreover, the preimage of each minimal open set U_y, is contractible, and then the restrictions $f|_{f^{-1}(U_y)} : f^{-1}(U_y) \to U_y$ are (weak) homotopy equivalences. Since the minimal basis is a basis like open cover, by Theorem 1.4.2, f is a weak homotopy equivalence. However, f is not a homotopy equivalence since its source and target are non homeomorphic minimal spaces.

Definition 1.4.4. Let X be a finite T_0-space. The *simplicial complex* $\mathcal{K}(X)$ *associated* to X (also called the *order complex*) is the simplicial complex whose simplices are the nonempty chains of X (see Fig. 1.3). Moreover, if $f : X \to Y$ is a continuous map between finite T_0-spaces, the *associated simplicial map* $\mathcal{K}(f) : \mathcal{K}(X) \to \mathcal{K}(Y)$ is defined by $\mathcal{K}(f)(x) = f(x)$.

Note that if $f : X \to Y$ is a continuous map between finite T_0-spaces, the vertex map $\mathcal{K}(f) : \mathcal{K}(X) \to \mathcal{K}(Y)$ is simplicial since f is order preserving and maps chains to chains.

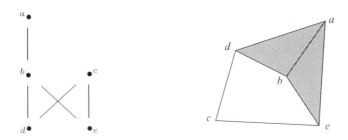

Fig. 1.3 A finite space and its associated simplicial complex

If X is a finite T_0-space, $\mathcal{K}(X) = \mathcal{K}(X^{op})$. Moreover, if $f : X \to Y$ is a continuous map between finite T_0-spaces, $\mathcal{K}(f) = \mathcal{K}(f^{op})$.

Let X be a finite T_0-space. A point α in the geometric realization $|\mathcal{K}(X)|$ of $\mathcal{K}(X)$ is a convex combination $\alpha = t_1 x_1 + t_2 x_2 + \ldots + t_r x_r$ where $\sum_{i=1}^{r} t_i = 1$, $t_i > 0$ for every $1 \le i \le r$ and $x_1 < x_2 < \ldots < x_r$ is a chain of X. The support or carrier of α is the set $support(\alpha) = \{x_1, x_2, \ldots, x_r\}$. We will see that the map $\alpha \mapsto x_1$ plays a fundamental role in this theory.

Definition 1.4.5. Let X be a finite T_0-space. Define the $\mathcal{K}$-*McCord map* $\mu_X : |\mathcal{K}(X)| \to X$ by $\mu_X(\alpha) = \min(support(\alpha))$.

The reader who is not familiar with some of the basic concepts about simplicial complexes that appear in the following proof, is suggested to see Appendix A.1.

Theorem 1.4.6. *The $\mathcal{K}$-McCord map μ_X is a weak homotopy equivalence for every finite T_0-space X.*

Proof. Notice that the minimal open sets U_x are contractible because they have maximum. We will prove that for each $x \in X$, $\mu_X^{-1}(U_x)$ is open and contractible. This will show that μ_X is continuous and that the restrictions $\mu_X|_{\mu_X^{-1}(U_x)} : \mu_X^{-1}(U_x) \to U_x$ are weak homotopy equivalences. Therefore, by Theorem 1.4.2, μ_X is a weak homotopy equivalence.

Let $x \in X$ and let $L = \mathcal{K}(X \smallsetminus U_x) \subseteq \mathcal{K}(X)$. In other words, L is the full subcomplex of K (possibly empty) spanned by the vertices which are not in U_x. We claim that

$$\mu_X^{-1}(U_x) = |\mathcal{K}(X)| \smallsetminus |L|.$$

If $\alpha \in \mu_X^{-1}(U_x)$, then $\min(support(\alpha)) \in U_x$. In particular, the support of α contains a vertex of U_x and then $\alpha \notin |L|$. Conversely, if $\alpha \notin |L|$, there exists $y \in support(\alpha)$ such that $y \in U_x$. Then $\min(support(\alpha)) \leq y \leq x$ and therefore $\mu_X(\alpha) \in U_x$. Since $|L| \subseteq |\mathcal{K}(X)|$ is closed, $\mu_X^{-1}(U_x)$ is open.

Now we show that $|\mathcal{K}(U_x)|$ is a strong deformation retract of $|\mathcal{K}(X)| \smallsetminus |L|$. This is a particular case of a more general fact. Let $i : |\mathcal{K}(U_x)| \hookrightarrow |\mathcal{K}(X)| \smallsetminus |L|$ be the inclusion. If $\alpha \in |\mathcal{K}(X)| \smallsetminus |L|$, $\alpha = t\beta + (1-t)\gamma$ for some $\beta \in |\mathcal{K}(U_x)|$, $\gamma \in |L|$ and $0 < t \leq 1$. Define $r : |\mathcal{K}(X)| \smallsetminus |L| \to |\mathcal{K}(U_x)|$ by $r(\alpha) = \beta$. Note that r is continuous since $r|_{(|\mathcal{K}(X)|\smallsetminus|L|)\cap\overline{\sigma}} : (|\mathcal{K}(X)| \smallsetminus |L|) \cap \overline{\sigma} \to \overline{\sigma}$ is continuous for every $\sigma \in \mathcal{K}(X)$. Here, $\overline{\sigma} \subseteq |\mathcal{K}(X)|$ denotes the closed simplex. Now, let $H : (|\mathcal{K}(X)|\smallsetminus|L|)\times I \to |\mathcal{K}(X)|\smallsetminus|L|$ be the linear homotopy between $1_{|\mathcal{K}(X)|\smallsetminus|L|}$ and ri, i.e.

$$H(\alpha, s) = (1 - s)\alpha + s\beta.$$

Then H is well defined and is continuous since each restriction

$$H|_{((|\mathcal{K}(X)|\smallsetminus|L|)\cap\overline{\sigma})\times I} : ((|\mathcal{K}(X)| \smallsetminus |L|) \cap \overline{\sigma}) \times I \to \overline{\sigma}$$

is continuous for every simplex σ of $\mathcal{K}(X)$. To prove the continuity of r and of H we use that $|\mathcal{K}(X)| \smallsetminus |L|$ has the final topology with respect to the subspaces $(|\mathcal{K}(X)| \smallsetminus |L|) \cap \overline{\sigma}$ for $\sigma \in \mathcal{K}(X)$. We could also argue that since the complexes involved are finite, the polyhedra have the metric topology and both r and H are continuous with respect to these metrics (see Appendix A.1).

Since every element of U_x is comparable with x, $\mathcal{K}(U_x)$ is a simplicial cone with apex x, that is, if σ is a simplex of $\mathcal{K}(U_x)$, then so is $\sigma \cup \{x\}$. In particular $|\mathcal{K}(U_x)|$ is contractible and then, so is $\mu_X^{-1}(U_x) = |\mathcal{K}(X)| \smallsetminus |L|$. $\square$

Remark 1.4.7. If $f : X \to Y$ is a continuous map between finite T_0-spaces, the following diagram commutes

$$
\begin{array}{ccc}
|\mathcal{K}(X)| & \xrightarrow{|\mathcal{K}(f)|} & |\mathcal{K}(Y)| \\
\downarrow{\mu_X} & & \downarrow{\mu_Y} \\
X & \xrightarrow{f} & Y
\end{array}
$$

since, for $\alpha \in |\mathcal{K}(X)|$,

$$f\mu_X(\alpha) = f(\min(support(\alpha))) = \min(f(support(\alpha)))$$
$$= \min(support(|\mathcal{K}(f)|(\alpha))) = \mu_Y|\mathcal{K}(f)|(\alpha).$$

Corollary 1.4.8. *Let $f : X \to Y$ be a map between finite T_0-spaces. Then f is a weak homotopy equivalence if and only if $|\mathcal{K}(f)| : |\mathcal{K}(X)| \to |\mathcal{K}(Y)|$ is a homotopy equivalence.*

Proof. Since μ_Y is a weak homotopy equivalence, by the 2-out-of-3 property, $|\mathcal{K}(f)|$ is a weak homotopy equivalence if and only if $\mu_Y|\mathcal{K}(f)| = f\mu_X$ is a weak homotopy equivalence. Since μ_X is a weak homotopy equivalence, this is equivalent to saying that f is a weak homotopy equivalence.   □

Corollary 1.4.9. *Let $f : X \to Y$ be a map between finite T_0-spaces. Then f is a weak homotopy equivalence if and only if f^{op} is a weak homotopy equivalence.*

Proof. Follows immediately from the previous result since $\mathcal{K}(f) = \mathcal{K}(f^{op})$.
                    □

Definition 1.4.10. Let K be a finite simplicial complex. The finite T_0-space $\mathcal{X}(K)$ *associated* to K (also called the *face poset* of K) is the poset of simplices of K ordered by inclusion. If $\varphi : K \to L$ is a simplicial map between finite simplicial complexes, there is a continuous map $\mathcal{X}(\varphi) : \mathcal{X}(K) \to \mathcal{X}(L)$ defined by $\mathcal{X}(\varphi)(\sigma) = \varphi(\sigma)$ for every simplex σ of K.

Example 1.4.11. If K is the 2-simplex, the associated finite space is the following

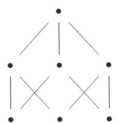

    If K is a finite complex, $\mathcal{K}(\mathcal{X}(K))$ is the first barycentric subdivision K' of K and if $\varphi : K \to L$ is a simplicial map, $\mathcal{K}(\mathcal{X}(\varphi)) = \varphi' : K' \to L'$ is the map induced in the barycentric subdivisions.
    Let $s_K : |K'| \to |K|$ be the linear homeomorphism defined by $s_K(\sigma) = b(\sigma)$ for every simplex σ of K. Here, $b(\sigma) \in |K|$ denotes the barycenter of σ. Define the $\mathcal{X}$-*McCord map* $\mu_K = \mu_{\mathcal{X}(K)}s_K^{-1} : |K| \to \mathcal{X}(K)$
    From 1.4.6 we deduce immediately the following result.

Theorem 1.4.12. *The $\mathcal{X}$-McCord map μ_K is a weak homotopy equivalence for every finite simplicial complex K.*

Proposition 1.4.13. *Let $\varphi : K \to L$ be a simplicial map between finite simplicial complexes. Then the following diagram commutes up to homotopy*

$$
\begin{array}{ccc}
|K| & \xrightarrow{\;|\varphi|\;} & |L| \\
\Big\downarrow{\scriptstyle \mu_K} & & \Big\downarrow{\scriptstyle \mu_L} \\
\mathcal{X}(K) & \xrightarrow{\;\mathcal{X}(\varphi)\;} & \mathcal{X}(L).
\end{array}
$$

Proof. Let $S = \{\sigma_1, \sigma_2, \ldots, \sigma_r\}$ be a simplex of K', where $\sigma_1 \subsetneq \sigma_2 \subsetneq \ldots \subsetneq \sigma_r$ is a chain of simplices of K. Let α be a point in the closed simplex $\overline{S}$. Then $s_K(\alpha) \in \overline{\sigma_r} \subseteq |K|$ and $|\varphi|s_K(\alpha) \in \overline{\varphi(\sigma_r)} \subseteq |L|$. On the other hand, $|\varphi'|(\alpha) \in \{\varphi(\sigma_1), \varphi(\sigma_2), \ldots, \varphi(\sigma_r)\}$ and then $s_L|\varphi'|(\alpha) \in \overline{\varphi(\sigma_r)}$. Therefore, the linear homotopy $H : |K'| \times I \to |L|$, $(\alpha, t) \mapsto (1-t)|\varphi|s_K(\alpha) + ts_L|\varphi'|(\alpha)$ is well defined and continuous. Then $|\varphi|s_K \simeq s_L|\varphi'|$ and, by Remark 1.4.7,

$$
\mu_L|\varphi| = \mu_{\mathcal{X}(L)}s_L^{-1}|\varphi| \simeq \mu_{\mathcal{X}(L)}|\varphi'|s_K^{-1}
$$

$$
= \mathcal{X}(\varphi)\mu_{\mathcal{X}(K)}s_K^{-1} = \mathcal{X}(\varphi)\mu_K.
$$

$\square$

Remark 1.4.14. An explicit homotopy between $\mu_L|\varphi|$ and $\mathcal{X}(\varphi)\mu_K$ is $\widetilde{H} = \mu_L H(s_K^{-1} \times 1_I)$. If $K_1 \subseteq K$ and $L_1 \subseteq L$ are subcomplexes and $\varphi(K_1) \subseteq L_1$ then $\widetilde{H}(|K_1| \times I) \subseteq \mathcal{X}(L_1) \subseteq \mathcal{X}(L)$.

From the 2-out-of-3 property and the fact that a map homotopic to a weak homotopy equivalence is also a weak homotopy equivalence, one deduces the following

Corollary 1.4.15. *Let* $\varphi : K \to L$ *be a simplicial map between finite simplicial complexes. Then* $|\varphi|$ *is a homotopy equivalence if and only if* $\mathcal{X}(\varphi) : \mathcal{X}(K) \to \mathcal{X}(L)$ *is a weak homotopy equivalence.*

From now on we will call *McCord maps* to both $\mathcal{K}$-McCord maps and $\mathcal{X}$-McCord maps, and it will be clear from the context which we are referring to.

Remark 1.4.16. As McCord explained in [55], the results of this section hold more generally for non-finite simplicial complexes and *A-spaces*. A topological space X is an *A-space* if arbitrary intersections of open subsets of X are again open. Finite spaces and moreover, locally finite spaces, are examples of *A*-spaces. If x is a point in an *A*-space X, the set U_x defined as above is also open. The correspondence between finite spaces and finite preordered sets trivially extends to a correspondence between *A*-spaces and preordered sets. If X is a T_0-*A*-space, the associated complex $\mathcal{K}(X)$ of finite chains is a well defined simplicial complex and the same proof of Theorem 1.4.6 shows that there is a weak homotopy equivalence $|\mathcal{K}(X)| \to X$. Conversely, given a simplicial complex K, the face poset $\mathcal{X}(K)$ is a locally finite space and there is a weak homotopy equivalence $|K| \to \mathcal{X}(K)$. Many of the results of this book can be stated in fact for A-spaces and general simplicial complexes. However, similarly to McCord's approach, we will focus our attention on the finite case. The first reason is that the deepest ideas and the complexity of the theory of A-spaces are already present in the finite setting. The second reason is that the most important applications of this theory, such as the study of minimal finite models (Chap. 3), simple homotopy types (Chap. 4), strong homotopy types (Chap. 5), Quillen's conjecture (Chap. 8) and the Andrews-Curtis conjecture (Chap. 11), can be analyzed or formulated only in the finite case.

Two topological spaces X and Y, not necessarily finite, are *weak homotopy equivalent* (or they are said to have the same *weak homotopy type*) if there exists a sequence of spaces $X = X_0, X_1, \ldots, X_n = Y$ such that there are weak homotopy equivalences $X_i \to X_{i+1}$ or $X_{i+1} \to X_i$ for every $0 \le i \le n-1$. Clearly this defines an equivalence relation. If two topological spaces X and Y are weak homotopy equivalent, we write $X \overset{we}{\approx} Y$. If X and Y are homotopy equivalent we write $X \overset{he}{\simeq} Y$ or just $X \simeq Y$.

If two topological spaces X and Y are weak homotopy equivalent, there exists a CW-complex Z (CW-approximation) and weak homotopy equivalences $Z \to X$ and $Z \to Y$ [38, Proposition 4.13, Corollary 4.19]. Two CW-complexes are weak homotopy equivalent if and only if they are homotopy equivalent. As we have seen, for finite spaces, weak homotopy equivalences are not in general homotopy equivalences. Moreover, there exist weak homotopy equivalent finite spaces such that there is no weak homotopy equivalence between them. However, if X and Y are two finite spaces which are weak homotopy equivalent, there exists a third finite space Z and weak homotopy equivalences $X \leftarrow Z \to Y$ (see Proposition 4.6.7).

Example 1.4.17. The non-Hausdorff suspension $\mathbb{S}(D_3)$ (see the paragraph below Definition 2.7.1) of the discrete space with three elements and its opposite $\mathbb{S}(D_3)^{op}$ have the same weak homotopy type, because there exist weak homotopy equivalences

$$\mathbb{S}(D_3) \leftarrow |\mathcal{K}(\mathbb{S}(D_3))| = |\mathcal{K}(\mathbb{S}(D_3)^{op})| \to \mathbb{S}(D_3)^{op}.$$

$$\mathbb{S}(D_3) \qquad\qquad\qquad \mathbb{S}(D_3)^{op}$$

However there is no weak homotopy equivalence between $\mathbb{S}(D_3)$ and $\mathbb{S}(D_3)^{op}$. In fact one can check that every map $\mathbb{S}(D_3) \to \mathbb{S}(D_3)^{op}$ factors through its image, which is a subspace of $\mathbb{S}(D_3)^{op}$ with trivial fundamental group or isomorphic to $\mathbb{Z}$. We exhibit a more elegant proof in Remark 3.4.9.

From Theorems 1.4.6 and 1.4.12 we immediately deduce the following result.

Corollary 1.4.18.

(a) Let X and Y be finite T_0-spaces. Then, $X \overset{we}{\approx} Y$ if and only if $|\mathcal{K}(X)| \overset{he}{\simeq} |\mathcal{K}(Y)|$.

(b) Let K and L be finite simplicial complexes. Then, $|K| \overset{he}{\simeq} |L|$ if and only if $\mathcal{X}(K) \overset{we}{\approx} \mathcal{X}(L)$.

McCord's Theorem 1.4.2 is one of the most useful tools to distinguish weak homotopy equivalences. Most of the times, we will apply this result to maps $f : X \to Y$ with Y finite, using the open cover given by the minimal basis of Y. Theorem 1.4.2 is closely related to the celebrated Quillen's Theorem A, which gives a sufficient condition for a functor between two categories to be a homotopy equivalence at the level of classifying spaces [69]. The so called Quillen's fiber Lemma [70, Proposition 1.6] is just Theorem A applied to the case in which both categories are finite posets and follows immediately from Theorem 1.4.2 also. It can be stated as follows.

Theorem 1.4.19 (McCord, Quillen). *Let $f : X \to Y$ be an order preserving map between finite posets such that $|\mathcal{K}(f^{-1}(U_y))|$ is contractible for every $y \in Y$. Then $|\mathcal{K}(f)| : |\mathcal{K}(X)| \to |\mathcal{K}(Y)|$ is a homotopy equivalence.*

Remark 1.4.20. In Quillen's paper [70] and in many other articles ([17, 18, 79, 80] for instance), posets are studied from a topological viewpoint only through their associated simplicial complexes. In some of those papers, when it is said that a finite poset is contractible it is meant that the associated polyhedron is contractible, and when an order preserving map is claimed to be a homotopy equivalence, this is regarded as the simplicial map between the associated complexes. In this book instead, finite posets are considered as topological spaces with an intrinsic topology. Although McCord's theory shows that a finite T_0-space and its order complex are closely related, it is not the same to say that a finite T_0-space X is contractible and that $|\mathcal{K}(X)|$ is contractible. A topological space X is said to be *homotopically trivial* or *weakly contractible* if all its homotopy groups are trivial. This is equivalent to saying that the map $X \to *$ is a weak homotopy equivalence. Since a finite T_0-space X and $|\mathcal{K}(X)|$ are weak homotopy equivalent, by the Whitehead Theorem it is equivalent to saying that X is homotopically trivial and that $|\mathcal{K}(X)|$ is contractible. However, the Whitehead Theorem does not hold for finite spaces and we will show explicit examples of homotopically trivial finite spaces which are not contractible (see Example 4.2.1). The contractibility of a finite T_0-space is equivalent to the combinatorial notion of *dismantlability* for posets [71]. The simplicial notion corresponding to the contractibility of a finite space is called *strong collapsibility* (see Corollary 5.2.8).

If $f : X \to Y$ is an order preserving map, then $|\mathcal{K}(f)| : |\mathcal{K}(X)| \to |\mathcal{K}(Y)|$ is a homotopy equivalence if and only if f is a weak homotopy equivalence between the finite spaces X and Y (Corollary 1.4.8). As we have seen, this is strictly weaker than f being a homotopy equivalence.

Some of the results which appear in this work could be restated in terms of the associated complexes, however our approach is inspired and motivated by McCord, Stong and May's topological viewpoint, which consists in considering finite topological spaces instead of posets and complexes.

We show now that Quillen's fiber Lemma follows from McCord's Theorem. The contractibility of $|\mathcal{K}(f^{-1}(U_y))|$ is equivalent to the fact that $f^{-1}(U_y)$ is homotopically trivial. Since U_y is contractible $f|_{f^{-1}(U_y)} : f^{-1}(U_y) \to U_y$ is a weak homotopy equivalence and then Theorem 1.4.2 says that $f : X \to Y$ is a weak homotopy equivalence. Therefore, $|\mathcal{K}(f)|$ is a homotopy equivalence by Corollary 1.4.8. Theorem 1.4.19 has simpler proofs than McCord's Theorem or Quillen's Theorem A. One, due to Walker [80], uses a homotopy version of the Acyclic carrier Theorem, and the other [6] was motivated by the ideas developed in this work. However, for our purposes Theorem 1.4.19 is not strong enough and McCord's Theorem will be needed in its general version.

The simplicial version of Quillen's Theorem A follows from the fiber Lemma and it states that if $\varphi : K \to L$ is a simplicial map and $|\varphi|^{-1}(\overline{\sigma})$ is contractible for every closed simplex $\overline{\sigma} \in |L|$, then $|\varphi|$ is a homotopy equivalence (see [69], p. 93).

Using this result, we prove a similar result to Theorem 1.4.2.

Proposition 1.4.21. *Let $f : X \to Y$ be a map between finite T_0-spaces such that $f^{-1}(c) \subseteq X$ is homotopically trivial for every chain c of Y. Then f is a weak homotopy equivalence.*

Proof. If c is a chain of Y or, equivalently, a simplex of $\mathcal{K}(Y)$, then $|\mathcal{K}(f)|^{-1}(\overline{c}) = |\mathcal{K}(f^{-1}(c))|$, which is contractible since $f^{-1}(c)$ is homotopically trivial. By Theorem A, $|\mathcal{K}(f)|$ is a homotopy equivalence and then f is a weak homotopy equivalence.   □

In fact, if the hypothesis of Proposition 1.4.21 holds, then $f^{-1}(U_y)$ is homotopically trivial for every $y \in Y$ and, by McCord Theorem, f is a weak homotopy equivalence. Therefore the proof of Proposition 1.4.21 is apparently superfluous. However, the proof of the first fact is a bit twisted, because it uses the very Proposition 1.4.21. If $f : X \to Y$ is such that $f^{-1}(c)$ is homotopically trivial for every chain c of Y, then each restriction $f|_{f^{-1}(U_y)} : f^{-1}(U_y) \to U_y$ satisfies the same hypothesis. Therefore, by Proposition 1.4.21, $f|_{f^{-1}(U_y)}$ is a weak homotopy equivalence and then $f^{-1}(U_y)$ is homotopically trivial.

In Sect. 4.4 we will prove, as an application of the simple homotopy theory of finite spaces, a simple homotopy version of Quillen's Theorem A for simplicial complexes.

Chapter 2
Basic Topological Properties
of Finite Spaces

In this chapter we present some results concerning elementary topological aspects of finite spaces. The proofs use basic elements of Algebraic Topology and have a strong combinatorial flavour. We study further homotopical properties including classical homotopy invariants and finite analogues of well-known topological constructions.

2.1   Homotopy and Contiguity

Recall that two simplicial maps $\varphi, \psi : K \to L$ are said to be *contiguous* if for every simplex $\sigma \in K$, $\varphi(\sigma) \cup \psi(\sigma)$ is a simplex of L. Two simplicial maps $\varphi, \psi : K \to L$ lie in the same *contiguity class* if there exists a sequence $\varphi = \varphi_0, \varphi_1, \ldots, \varphi_n = \psi$ such that φ_i and φ_{i+1} are contiguous for every $0 \le i < n$.

If $\varphi, \psi : K \to L$ lie in the same contiguity class, the induced maps in the geometric realizations $|\varphi|, |\psi| : |K| \to |L|$ are homotopic (see Corollary A.1.3 of the appendix).

In this section we study the relationship between contiguity classes of simplicial maps and homotopy classes of the associated maps between finite spaces. These results appear in [11].

Lemma 2.1.1. *Let $f, g : X \to Y$ be two homotopic maps between finite T_0-spaces. Then there exists a sequence $f = f_0, f_1, \ldots, f_n = g$ such that for every $0 \le i < n$ there is a point $x_i \in X$ with the following properties:*

1. f_i and f_{i+1} coincide in $X \smallsetminus \{x_i\}$, and
2. $f_i(x_i) \prec f_{i+1}(x_i)$ or $f_{i+1}(x_i) \prec f_i(x_i)$.

Proof. Without loss of generality, we may assume that $f = f_0 \le g$ by Corollary 1.2.6. Let $A = \{x \in X \mid f(x) \ne g(x)\}$. If $A = \emptyset$, $f = g$ and there is nothing to prove. Suppose $A \ne \emptyset$ and let $x = x_0$ be a maximal point

J.A. Barmak, *Algebraic Topology of Finite Topological Spaces and Applications*, Lecture Notes in Mathematics 2032,
DOI 10.1007/978-3-642-22003-6_2, © Springer-Verlag Berlin Heidelberg 2011

of A. Let $y \in Y$ be such that $f(x) \prec y \le g(x)$ and define $f_1 : X \to Y$ by $f_1|_{X \smallsetminus \{x\}} = f|_{X \smallsetminus \{x\}}$ and $f_1(x) = y$. Then f_1 is continuous for if $x' > x$, $x' \notin A$ and therefore

$$f_1(x') = f(x') = g(x') \ge g(x) \ge y = f_1(x).$$

Repeating this construction for f_i and g, we define f_{i+1}. By finiteness of X and Y this process ends. $\square$

Proposition 2.1.2. *Let $f, g : X \to Y$ be two homotopic maps between finite T_0-spaces. Then the simplicial maps $\mathcal{K}(f), \mathcal{K}(g) : \mathcal{K}(X) \to \mathcal{K}(Y)$ lie in the same contiguity class.*

Proof. By the previous lemma, we can assume that there exists $x \in X$ such that $f(y) = g(y)$ for every $y \ne x$ and $f(x) \prec g(x)$. Therefore, if C is a chain in X, $f(C) \cup g(C)$ is a chain on Y. In other words, if $\sigma \in \mathcal{K}(X)$ is a simplex, $\mathcal{K}(f)(\sigma) \cup \mathcal{K}(g)(\sigma)$ is a simplex in $\mathcal{K}(Y)$. $\square$

Proposition 2.1.3. *Let $\varphi, \psi : K \to L$ be simplicial maps which lie in the same contiguity class. Then $\mathcal{X}(\varphi) \simeq \mathcal{X}(\psi)$.*

Proof. Assume that φ and ψ are contiguous. Then the map $f : \mathcal{X}(K) \to \mathcal{X}(L)$, defined by $f(\sigma) = \varphi(\sigma) \cup \psi(\sigma)$ is well-defined and continuous. Moreover $\mathcal{X}(\varphi) \le f \ge \mathcal{X}(\psi)$, and then $\mathcal{X}(\varphi) \simeq \mathcal{X}(\psi)$. $\square$

2.2 Minimal Pairs

In this section we generalize Stong's ideas on homotopy types to the case of pairs (X, A) of finite spaces (i.e. a finite space X and a subspace $A \subseteq X$). As a consequence, we will deduce that every core of a finite T_0-space can be obtained by removing beat points from X. Here we introduce the notion of *strong collapse* which plays a central role in Chap. 5. Most of the results of this section appear in [11].

Definition 2.2.1. A pair (X, A) of finite T_0-spaces is a *minimal pair* if all the beat points of X are in A.

The next result generalizes the result of Stong (the case $A = \emptyset$) studied in Sect. 1.3 and its proof is very similar to the original one.

Proposition 2.2.2. *Let (X, A) be a minimal pair and let $f : X \to X$ be a map such that $f \simeq 1_X$ rel A. Then $f = 1_X$.*

Proof. Suppose that $f \le 1_X$ and $f|_A = 1_A$. Let $x \in X$. If $x \in X$ is minimal, $f(x) = x$. In general, suppose we have proved that $f|_{\hat{U}_x} = 1|_{\hat{U}_x}$. If $x \in A$, $f(x) = x$. If $x \notin A$, x is not a down beat point of X. However $y < x$ implies $y = f(y) \le f(x) \le x$. Therefore $f(x) = x$. The case $f \ge 1_X$ is similar, and the general case follows from Corollary 1.2.6. $\square$

Corollary 2.2.3. *Let* (X, A) *and* (Y, B) *be minimal pairs,* $f : X \to Y$, $g : Y \to X$ *such that* $gf \simeq 1_X$ *rel* A, $gf \simeq 1_Y$ *rel* B. *Then* f *and* g *are homeomorphisms.*

Definition 2.2.4. If x is a beat point of a finite T_0-space X, we say that there is an *elementary strong collapse* from X to $X \smallsetminus x$ and write $X \searrow^e X \smallsetminus x$. There is a *strong collapse* $X \searrow Y$ (or a *strong expansion* $Y \nearrow X$) if there is a sequence of elementary strong collapses starting in X and ending in Y.

Stong's results show that two finite T_0-spaces are homotopy equivalent if and only if there exists a sequence of strong collapses and strong expansions from X to Y (since the later is true for homeomorphic spaces).

Corollary 2.2.5. *Let* X *be a finite* T_0-*space and let* $A \subseteq X$. *Then,* $X \searrow A$ *if and only if* A *is a strong deformation retract of* X.

Proof. If $X \searrow A$, $A \subseteq X$ is a strong deformation retract. This was already proved by Stong (see Sect. 1.3). Conversely, suppose $A \subseteq X$ is a strong deformation retract. Perform arbitrary elementary strong collapses removing beat points which are not in A. Suppose $X \searrow Y \supseteq A$ and that all the beat points of Y lie in A. Then (Y, A) is a minimal pair. Since A and Y are strong deformation retracts of X, the minimal pairs (A, A) and (Y, A) are in the hypothesis of Corollary 2.2.3. Therefore A and Y are homeomorphic and so, $X \searrow Y = A$. $\qquad\square$

Example 2.2.6. The space X

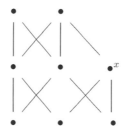

is contractible, but the point x is not a strong deformation retract of X, because $(X, \{x\})$ is a minimal pair.

Corollary 2.2.7. *Let* (X, A) *be a minimal pair such that* A *is a minimal finite space and* $f \simeq 1_{(X,A)} : (X, A) \to (X, A)$. *Then* $f = 1_X$.

If X and Y are homotopy equivalent finite T_0-spaces, the associated polyhedra $|\mathcal{K}(X)|$ and $|\mathcal{K}(Y)|$ also have the same homotopy type. However the converse is obviously false, since the associated polyhedra are homotopy equivalent if and only if the finite spaces are weak homotopy equivalent.

In Chap. 5 we will study the notion of *strong homotopy types* of simplicial complexes which have a very simple description and corresponds exactly to the concept of homotopy types of the associated finite spaces.

2.3 T_1-Spaces

We will prove that Hausdorff spaces do not have in general the homotopy type of any finite space. Recall that a topological space X satisfies the T_1-separation axiom if for any two distinct points $x, y \in X$ there exist open sets U and V such that $x \in U$, $y \in V$, $y \notin U$, $x \notin V$. This is equivalent to saying that the points are closed in X. All Hausdorff spaces are T_1, but the converse is false.

If a finite space is T_1, then every subset is closed and so, X is discrete.

Since the core X_c of a finite space X is the disjoint union of the cores of its connected components, we can deduce the following

Lemma 2.3.1. *Let X be a finite space such that X_c is discrete. Then X is a disjoint union of contractible spaces.*

Theorem 2.3.2. *Let X be a finite space and let Y be a T_1-space homotopy equivalent to X. Then X is a disjoint union of contractible spaces.*

Proof. Since $X \simeq Y$, $X_c \simeq Y$. Let $f : X_c \to Y$ be a homotopy equivalence with homotopy inverse g. Then $gf = 1_{X_c}$ by Theorem 1.3.6. Since f is a one to one map from X_c to a T_1-space, it follows that X_c is also T_1 and therefore discrete. Now the result follows from the previous lemma. $\square$

Remark 2.3.3. The proof of the previous theorem can be done without using Theorem 1.3.6, showing that any map $f : X \to Y$ from a finite space to a T_1-space must be locally constant.

Corollary 2.3.4. *Let Y be a connected and non contractible T_1-space. Then Y does not have the same homotopy type as any finite space.*

Proof. Follows immediately from Theorem 2.3.2. $\square$

For example, for any $n \geq 1$, the n-dimensional sphere S^n does not have the homotopy type of any finite space. However, S^n does have, as any finite polyhedron, the same weak homotopy type as some finite space.

2.4   Loops in the Hasse Diagram and the Fundamental Group

In this section we give a full description of the fundamental group of a finite T_0-space in terms of its Hasse diagram. This characterization is induced from the well known description of the fundamental group of a simplicial complex. The Hasse diagram of a finite T_0-space X will be denoted $\mathcal{H}(X)$, and $\mathrm{E}(\mathcal{H}(X))$ will denote the set of edges of the digraph $\mathcal{H}(X)$.

Recall that an *edge-path* in a simplicial complex K, is a sequence (v_0, v_1), $(v_1, v_2), \ldots, (v_{r-1}, v_r)$ of ordered pairs of vertices in which $\{v_i, v_{i+1}\}$ is a simplex for every i. If an edge-path contains two consecutive pairs (v_i, v_{i+1}), (v_{i+1}, v_{i+2}) where $\{v_i, v_{i+1}, v_{i+2}\}$ is a simplex, we can replace the two pairs by a unique pair (v_i, v_{i+2}) to obtain an *equivalent* edge-path. The equivalence classes of edge-paths are the morphisms of a groupoid called the *edge-path groupoid* of K, which is denoted by $E(K)$. The full subcategory of edge-paths with origin and end v_0 is the *edge-path group* $E(K, v_0)$ which is isomorphic to the fundamental group $\pi_1(|K|, v_0)$ (see [75, Sect. 3.6] for more details).

Definition 2.4.1. Let (X, x_0) be a finite pointed T_0-space. An ordered pair of points $e = (x, y)$ is called an $\mathcal{H}$-*edge* of X if $(x, y) \in E(\mathcal{H}(X))$ or $(y, x) \in E(\mathcal{H}(X))$. The point x is called the *origin* of e and denoted $x = \mathfrak{o}(e)$, the point y is called the *end* of e and denoted $y = \mathfrak{e}(e)$. The *inverse* of an $\mathcal{H}$-edge $e = (x, y)$ is the $\mathcal{H}$-edge $e^{-1} = (y, x)$.

An $\mathcal{H}$-*path* in (X, x_0) is a finite sequence (possibly empty) of $\mathcal{H}$-edges $\xi = e_1 e_2 \ldots e_n$ such that $\mathfrak{e}(e_i) = \mathfrak{o}(e_{i+1})$ for all $1 \le i \le n-1$. The *origin* of a non empty $\mathcal{H}$-path ξ is $\mathfrak{o}(\xi) = \mathfrak{o}(e_1)$ and its *end* is $\mathfrak{e}(\xi) = \mathfrak{e}(e_n)$. The origin and the end of the empty $\mathcal{H}$-path is $\mathfrak{o}(\emptyset) = \mathfrak{e}(\emptyset) = x_0$. If $\xi = e_1 e_2 \ldots e_n$, we define $\bar{\xi} = e_n^{-1} e_{n-1}^{-1} \ldots e_1^{-1}$. If ξ, ξ' are $\mathcal{H}$-paths such that $\mathfrak{e}(\xi) = \mathfrak{o}(\xi')$, we define the product $\mathcal{H}$-path $\xi\xi'$ as the concatenation of the sequence ξ followed by the sequence ξ'.

An $\mathcal{H}$-path $\xi = e_1 e_2 \ldots e_n$ is said to be *monotonic* if $e_i \in E(\mathcal{H}(X))$ for all $1 \le i \le n$ or $e_i^{-1} \in E(\mathcal{H}(X))$ for all $1 \le i \le n$.

A *loop* at x_0 is an $\mathcal{H}$-path that starts and ends in x_0. Given two loops ξ, ξ' at x_0, we say that they are *close* if there exist $\mathcal{H}$-paths $\xi_1, \xi_2, \xi_3, \xi_4$ such that ξ_2 and ξ_3 are monotonic and the set $\{\xi, \xi'\}$ coincides with $\{\xi_1\xi_2\xi_3\xi_4, \xi_1\xi_4\}$.

We say that two loops ξ, ξ' at x_0 are $\mathcal{H}$-*equivalent* if there exists a finite sequence of loops $\xi = \xi_1, \xi_2, \ldots, \xi_n = \xi'$ such that any two consecutive are close. We denote by $\langle \xi \rangle$ the $\mathcal{H}$-equivalence class of a loop ξ and $\mathscr{H}(X, x_0)$ the set of these classes.

Theorem 2.4.2. *Let (X, x_0) be a pointed finite T_0-space. Then the product $\langle \xi \rangle \langle \xi' \rangle = \langle \xi\xi' \rangle$ is well defined and induces a group structure on $\mathscr{H}(X, x_0)$.*

Proof. It is easy to check that the product is well defined, associative and that $\langle \emptyset \rangle$ is the identity. In order to prove that the inverse of $\langle e_1 e_2 \ldots e_n \rangle$ is $\langle e_n^{-1} e_{n-1}^{-1} \ldots e_1^{-1} \rangle$ we need to show that for any composable $\mathcal{H}$-paths ξ, ξ' such that $\mathfrak{o}(\xi) = \mathfrak{e}(\xi') = x_0$ and for any $\mathcal{H}$-edge e, composable with ξ, one has that $\langle \xi e e^{-1} \xi' \rangle = \langle \xi\xi' \rangle$. But this follows immediately from the definition of close loops since e and e^{-1} are monotonic. $\qquad\square$

Theorem 2.4.3. *Let (X, x_0) be a pointed finite T_0-space. Then the edge-path group $E(\mathcal{K}(X), x_0)$ of $\mathcal{K}(X)$ with base vertex x_0 is isomorphic to $\mathscr{H}(X, x_0)$.*

Proof. Let us define

$$\varphi : \mathscr{H}(X, x_0) \longrightarrow E(\mathcal{K}(X), x_0),$$

$$\langle e_1 e_2 \ldots e_n \rangle \longmapsto [e_1 e_2 \ldots e_n],$$

$$\langle \emptyset \rangle \longmapsto [(x_0, x_0)],$$

where $[\xi]$ denotes the class of ξ in $E(\mathcal{K}(X), x_0)$.

To prove that φ is well defined, let us suppose that the loops $\xi_1 \xi_2 \xi_3 \xi_4$ and $\xi_1 \xi_4$ are close, where $\xi_2 = e_1 e_2 \ldots e_n$, $\xi_3 = e_1' e_2' \ldots e_m'$ are monotonic $\mathcal{H}$-paths. By induction, it can be proved that

$$[\xi_1 \xi_2 \xi_3 \xi_4] = [\xi_1 e_1 e_2 \ldots e_{n-j}(\mathfrak{o}(e_{n-j+1}), \mathfrak{e}(e_n))\xi_3 \xi_4]$$

for $1 \le j \le n$. In particular $[\xi_1 \xi_2 \xi_3 \xi_4] = [\xi_1(\mathfrak{e}(\xi_1), \mathfrak{e}(e_n))\xi_3 \xi_4]$.

Analogously,

$$[\xi_1(\mathfrak{e}(\xi_1), \mathfrak{e}(e_n))\xi_3 \xi_4] = [\xi_1(\mathfrak{e}(\xi_1), \mathfrak{e}(e_n))(\mathfrak{o}(e_1'), \mathfrak{o}(\xi_4))\xi_4]$$

and then

$$[\xi_1 \xi_2 \xi_3 \xi_4] = [\xi_1(\mathfrak{e}(\xi_1), \mathfrak{e}(e_n))(\mathfrak{o}(e_1'), \mathfrak{o}(\xi_4))\xi_4]$$

$$= [\xi_1(\mathfrak{e}(\xi_1), \mathfrak{e}(e_n))(\mathfrak{e}(e_n), \mathfrak{e}(\xi_1))\xi_4] = [\xi_1(\mathfrak{e}(\xi_1), \mathfrak{e}(\xi_1))\xi_4] = [\xi_1 \xi_4].$$

If $\xi = (x_0, x_1)(x_1, x_2) \ldots (x_{n-1}, x_n)$ is an edge-path in $\mathcal{K}(X)$ with $x_n = x_0$, then x_{i-1} and x_i are comparable for all $1 \le i \le n$. In this case, we can find monotonic $\mathcal{H}$-paths $\xi_1, \xi_2, \ldots, \xi_n$ such that $\mathfrak{o}(\xi_i) = x_{i-1}$, $\mathfrak{e}(\xi_i) = x_i$ for all $1 \le i \le n$. Let us define

$$\psi : E(\mathcal{K}(X), x_0) \longrightarrow \mathscr{H}(X, x_0),$$

$$[\xi] \longmapsto \langle \xi_1 \xi_2 \ldots \xi_n \rangle.$$

This definition does not depend on the choice of the $\mathcal{H}$-paths ξ_i since if two choices differ only for $i = k$ then $\xi_1 \ldots \xi_k \ldots \xi_n$ and $\xi_1 \ldots \xi_k' \ldots \xi_n$ are $\mathcal{H}$-equivalent because both of them are close to $\xi_1 \ldots \xi_k \xi_k^{-1} \xi_k' \ldots \xi_n$.

The definition of ψ does not depend on the representative. Suppose that $\xi'(x, y)(y, z)\xi''$ and $\xi'(x, z)\xi''$ are simply equivalent edge-paths in $\mathcal{K}(X)$ that start and end in x_0, where ξ and ξ' are edge-paths and x, y, z are comparable. In the case that y lies between x and z, we can choose the monotonic $\mathcal{H}$-path corresponding to (x, z) to be the juxtaposition of the corresponding to (x, y) and (y, z), and so ψ is equally defined in both edge-paths. In the case that $z \le x \le y$ we can choose monotonic $\mathcal{H}$-paths α, β from x to y and from z to x, and then α will be the corresponding $\mathcal{H}$-path to (x, y), $\overline{\alpha}\beta$ that corresponding to (y, z) and $\overline{\beta}$ to (x, z). It only remains to prove that $\langle \gamma' \alpha \overline{\alpha} \beta \gamma'' \rangle = \langle \gamma' \overline{\beta} \gamma'' \rangle$ for $\mathcal{H}$-paths γ' and γ'', which is trivial. The other cases are analogous to the last one.

It is clear that φ and ψ are mutually inverse.   □

Since $E(\mathcal{K}(X), x_0)$ is isomorphic to $\pi_1(|\mathcal{K}(X)|, x_0)$ (cf. [75, Corollary 3.6.17]), we obtain the following result.

Corollary 2.4.4. *Let (X, x_0) be a pointed finite T_0-space, then $\mathscr{H}(X, x_0) = \pi_1(X, x_0)$.*

Remark 2.4.5. Since every finite space is homotopy equivalent to a finite T_0-space, this computation of the fundamental group can be applied to any finite space.

2.5 Euler Characteristic

If the homology (with integer coefficients) of a topological space X is finitely generated as a graded abelian group, the Euler characteristic of X is defined by $\chi(X) = \sum_{n \geq 0} (-1)^n rank(H_n(X))$. If Z is a compact CW-complex, its homology is finitely generated and $\chi(Z) = \sum_{n \geq 0} (-1)^n \alpha_n$ where α_n is the number of n-cells of Z. A weak homotopy equivalence induces isomorphisms in homology groups and therefore weak homotopy equivalent spaces have the same Euler characteristic.

Since any finite T_0-space X is weak homotopy equivalent to the geometric realization of $\mathcal{K}(X)$, whose simplices are the non empty chains of X, the Euler characteristic of X is

$$\chi(X) = \sum_{C \in \mathcal{C}(X)} (-1)^{\#C+1}, \tag{2.1}$$

where $\mathcal{C}(X)$ is the set of nonempty chains of X and $\#C$ is the cardinality of C.

We will give a basic combinatorial proof of the fact that the Euler characteristic is a homotopy invariant in the setting of finite spaces, using only the formula 2.1 as definition.

Theorem 2.5.1. *Let X and Y be finite T_0-spaces with the same homotopy type. Then $\chi(X) = \chi(Y)$.*

Proof. Let X_c and Y_c be cores of X and Y. Then there exist two sequences of finite T_0-spaces $X = X_0 \supseteq \ldots \supseteq X_n = X_c$ and $Y = Y_0 \supseteq \ldots \supseteq Y_m = Y_c$, where X_{i+1} is constructed from X_i by removing a beat point and Y_{i+1} is constructed from Y_i, similarly. Since X and Y are homotopy equivalent, X_c and Y_c are homeomorphic. Thus, $\chi(X_c) = \chi(Y_c)$.

It suffices to show that the Euler characteristic does not change when a beat point is removed. Let P be a finite poset and let $p \in P$ be a beat point. Then there exists $q \in P$ such that if r is comparable with p then r is comparable with q.

Hence we have a bijection

$$\varphi : \{C \in \mathcal{C}(P) \mid p \in C,\ q \notin C\} \longrightarrow \{C \in \mathcal{C}(P) \mid p \in C,\ q \in C\},$$
$$C \longmapsto C \cup \{q\}.$$

Therefore

$$\chi(P) - \chi(P \smallsetminus \{p\}) = \sum_{p \in C \in \mathcal{C}P} (-1)^{\#C+1} = \sum_{q \notin C \ni p} (-1)^{\#C+1} + \sum_{q \in C \ni p} (-1)^{\#C+1}$$

$$= \sum_{q \notin C \ni p} (-1)^{\#C+1} + \sum_{q \notin C \ni p} (-1)^{\#\varphi(C)+1} = \sum_{q \notin C \ni p} (-1)^{\#C+1} + \sum_{q \notin C \ni p} (-1)^{\#C} = 0.$$

$\square$

The Euler characteristic of finite T_0-spaces is intimately related to the Möbius function of posets, which is a generalization of the classical Möbius function of number theory. We will say just a few words about this. For proofs and applications we refer the reader to [29].

Given a finite poset P, we define the *incidence algebra* $\mathfrak{A}(P)$ of P as the set of functions $P \times P \to \mathbb{R}$ such that $f(x,y) = 0$ if $x \not\leq y$ with the usual structure of $\mathbb{R}$-vector space and the product given by

$$fg(x,y) = \sum_{z \in P} f(x,z)g(z,y).$$

The element $\zeta_P \in \mathfrak{A}(P)$ defined by $\zeta_P(x,y) = 1$ if $x \leq y$ and 0 in other case, is invertible in $\mathfrak{A}(P)$. The *Möbius fuction* $\mu_P \in \mathfrak{A}(P)$ is the inverse of ζ_P.

The Theorem of Hall states that if P is a finite poset and $x, y \in P$, then $\mu_P(x,y) = \sum_{n \geq 0} (-1)^{n+1} c_n$, where c_n is the number of chains of n-elements which start in x and end in y.

Given a finite poset P, $\hat{P} = P \cup \{0,1\}$ denotes the poset obtained when adjoining a minimum 0 and a maximum 1 to P. In particular, (2.1) and the Theorem of Hall, give the following

Corollary 2.5.2. *Let P be a finite poset. Then*

$$\widetilde{\chi}(P) = \mu_{\hat{P}}(0,1),$$

where $\widetilde{\chi}(P) = \chi(P) - 1$ denotes the reduced Euler characteristic of the finite space P.

One of the motivations of the Möbius function is the following inversion formula.

Theorem 2.5.3 (Möbius inversion formula). *Let P be a finite poset and let $f, g : P \to \mathbb{R}$. Then*

$$g(x) = \sum_{y \leq x} f(y) \ \text{if and only if} \ f(x) = \sum_{y \leq x} \mu_P(y, x) g(y).$$

Analogously,

$$g(x) = \sum_{y \geq x} f(y) \ \text{if and only if} \ f(x) = \sum_{y \geq x} \mu_P(y, x) g(y).$$

Beautiful applications of these formulae are: (1) the Möbius inversion of number theory which is obtained when applying Theorem 2.5.3 to the order given by divisibility of the integer numbers; (2) the inclusion–exclusion formula obtained from the power set of a set ordered by inclusion.

2.6 Automorphism Groups of Finite Posets

It is well known that any finite group G can be realized as the automorphism group of a finite poset. In 1946 Birkhoff [13] proved that if the order of G is n, G can be realized as the automorphisms of a poset with $n(n+1)$ points. In 1972 Thornton [78] improved slightly Birkhoff's result: He obtained a poset of $n(2r + 1)$ points, when the group is generated by r elements.

We present here a result which appears in [10]. Following Birkhoff's and Thornton's ideas, we exhibit a simple proof of the following fact which improves their results

Theorem 2.6.1. *Given a group G of finite order n with r generators, there exists a poset X with $n(r + 2)$ points such that $Aut(X) \simeq G$.*

Recall first that the *height* $ht(X)$ of a finite poset X is one less than the maximum number of elements in a chain of X. The *height* of a point x in a finite poset X is $ht(x) = ht(U_x)$.

Proof. Let $\{h_1, h_2, \ldots, h_r\}$ be a set of r generators of G. We define the poset $X = G \times \{-1, 0, \ldots, r\}$ with the following order

- $(g, i) \leq (g, j)$ if $-1 \leq i \leq j \leq r$
- $(gh_i, -1) \leq (g, j)$ if $1 \leq i \leq j \leq r$

Define $\phi : G \to Aut(X)$ by $\phi(g)(h, i) = (gh, i)$. It is easy to see that $\phi(g) : X \to X$ is order preserving and that it is an automorphism with inverse $\phi(g^{-1})$. Therefore ϕ is a well defined homomorphism. Clearly ϕ is a monomorphism since $\phi(g) = 1$ implies $(g, -1) = \phi(g)(e, -1) = (e, -1)$.

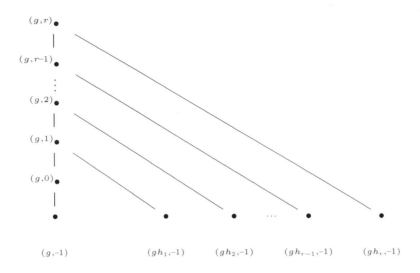

Fig. 2.1 $U_{(g,r)}$

It remains to show that ϕ is an epimorphism. Let $f : X \to X$ be an automorphism. Since $(e, -1)$ is minimal in X, so is $f(e, -1)$ and therefore $f(e, -1) = (g, -1)$ for some $g \in G$. We will prove that $f = \phi(g)$.

Let $Y = \{x \in X \mid f(x) = \phi(g)(x)\}$. Y is nonempty since $(e, -1) \in Y$. We prove first that Y is an open subspace of X. Suppose $x = (h, i) \in Y$. Then the restrictions

$$f|_{U_x}, \phi(g)|_{U_x} : U_x \to U_{f(x)}$$

are isomorphisms. On the other hand, there exists a unique automorphism $U_x \to U_x$ since the unique chain of $i + 2$ elements must be fixed by any such automorphism. Thus, $f|_{U_x}^{-1}\phi(g)|_{U_x} = 1_{U_x}$, and then $f|_{U_x} = \phi(g)|_{U_x}$, which proves that $U_x \subseteq Y$. Similarly we see that $Y \subseteq X$ is closed. Assume $x = (h, i) \notin Y$. Since $f \in Aut(X)$, it preserves the height of any point. In particular $ht(f(x)) = ht(x) = i+1$ and therefore $f(x) = (k, i) = \phi(kh^{-1})(x)$ for some $k \in G$. Moreover $k \neq gh$ since $x \notin Y$. As above, $f|_{U_x} = \phi(kh^{-1})|_{U_x}$, and since $kh^{-1} \neq g$ we conclude that $U_x \cap Y = \emptyset$.

We prove now that X is connected. It suffices to prove that any two minimal elements of X are in the same connected component. Given $h, k \in G$, we have $h = kh_{i_1}h_{i_2}\ldots h_{i_m}$ for some $1 \leq i_1, i_2\ldots i_m \leq r$. On the other hand, $(kh_{i_1}h_{i_2}\ldots h_{i_s}, -1)$ and $(kh_{i_1}h_{i_2}\ldots h_{i_{s+1}}, -1)$ are connected via $(kh_{i_1}h_{i_2}\ldots h_{i_s}, -1) < (kh_{i_1}h_{i_2}\ldots h_{i_s}, r) > (kh_{i_1}h_{i_2}\ldots h_{i_{s+1}}, -1)$. This implies that $(k, -1)$ and $(h, -1)$ are in the same connected component.

Finally, since X is connected and Y is closed, open and nonempty, $Y = X$, i.e. $f = \phi(g)$. Therefore ϕ is an epimorphism, and then $G \simeq Aut(X)$. $\square$

If the generators $h_1, h_2, \ldots, h_r$ are non-trivial, the open sets $U_{(g,r)}$ are as in Fig. 2.1. In that case it is not hard to prove that the finite space X constructed above is weak homotopy equivalent to a wedge of $n(r-1)+1$ circles, or in other words, that the order complex of X is homotopy equivalent to a wedge of $n(r-1)+1$ circles. The space X deformation retracts to the subspace $Y = G \times \{-1, r\}$ of its minimal and maximal points. A retraction is given by the map $f : X \to Y$, defined as $f(g, i) = (g, r)$ if $i \geq 0$ and $f(g, -1) = (g, -1)$. Now the order complex $\mathcal{K}(Y)$ of Y is a connected simplicial complex of dimension 1, so its homotopy type is completely determined by its Euler Characteristic. This complex has $2n$ vertices and $n(r+1)$ edges, which means that it has the homotopy type of a wedge of $1 - \chi(\mathcal{K}(Y)) = n(r-1)+1$ circles.

On the other hand, note that in general the automorphism group of a finite space, does not say much about its homotopy type as we see in the following

Proposition 2.6.2. *Given a finite group G and a finite space X, there exists a finite space Y which is homotopy equivalent to X and such that $Aut(Y) \simeq G$.*

Proof. We make this construction in two steps. First, we find a finite T_0-space $\tilde{X}$ homotopy equivalent to X and such that $Aut(\tilde{X}) = 0$. To do this, assume that X is T_0 and consider a linear extension $x_1, x_2, \ldots, x_n$ of the poset X (i.e. $X = \{x_1, x_2, \ldots, x_n\}$ and $x_i \leq x_j$ implies $i \leq j$). Now, for each $1 \leq k \leq n$ attach a chain of length kn to X with minimum x_{n-k+1}. The resulting space $\tilde{X}$ deformation retracts to X and every automorphism $f : \tilde{X} \to \tilde{X}$ must fix the unique chain C_1 of length n^2 (with minimum x_1). Therefore f restricts to a homeomorphism $\tilde{X} \smallsetminus C_1 \to \tilde{X} \smallsetminus C_1$ which must fix the unique chain C_2 of length $n(n-1)$ of $\tilde{X} \smallsetminus C_1$ (with minimum x_2). Applying this reasoning repeatedly, we conclude that f fixes every point of $\tilde{X}$. On the other hand, we know that there exists a finite T_0-space Z such that $Aut(Z) = G$.

Now the space Y is constructed as follows. It contains one copy of $\tilde{X}$ and one of Z, and the additional relations $z \leq x$ for every $z \in Z$ and $x \geq x_1$ in $\tilde{X}$. So, all the elements of Z are smaller than $x_1 \in \tilde{X}$. Clearly Y deformation retracts to $\tilde{X}$. Moreover, if $f : Y \to Y$ is an automorphism, $f(x_1) \notin Z$ since $f(x_1)$ cannot be comparable with x_1 and distinct from it (cf. Lemma 8.1.1). Since there is only one chain of n^2 elements in $\tilde{X}$, it must be fixed by f. In particular $f(x_1) = x_1$, and then $f|_Z : Z \to Z$. Thus f restricts to automorphisms of $\tilde{X}$ and of Z and therefore $Aut(Y) \simeq Aut(Z) \simeq G$. $\square$

2.7   Joins, Products, Quotients and Wedges

In this section we will study some basic constructions in the settings of finite spaces, simplicial complexes and general topological spaces. We will relate these constructions to each other and analyze them from the homotopical point of view.

Recall that the *simplicial join* $K * L$ (also denoted by KL) of two simplicial complexes K and L (with disjoint vertex sets) is the complex

$$K * L = K \cup L \cup \{\sigma \cup \tau | \ \sigma \in K, \tau \in L\}.$$

The simplicial cone aK with base K is the join of K with a vertex $a \notin K$. It is well known that for finite simplicial complexes K and L, the geometric realization $|K * L|$ is homeomorphic to the topological join $|K| * |L|$. If K is the 0-complex with two vertices, $|K * L| = |K| * |L| = S^0 * |L| = \Sigma|L|$ is the suspension of $|L|$. Here, S^0 denotes the discrete space on two points (0-sphere).

There is an analogous construction for finite spaces.

Definition 2.7.1. The *(non-Hausdorff) join* (also called the *ordinal sum*) $X \circledast Y$ of two finite T_0-spaces X and Y is the disjoint union $X \sqcup Y$ keeping the given ordering within X and Y and setting $x \leq y$ for every $x \in X$ and $y \in Y$.

Note that the join is associative and in general $X \circledast Y \neq Y \circledast X$. Special cases of joins are the *non-Hausdorff cone* $\mathbb{C}(X) = X \circledast D^0$ and the *non-Hausdorff suspension* $\mathbb{S}(X) = X \circledast S^0$ of any finite T_0-space X. Here $D^0 = *$ denotes the singleton (0-cell).

Remark 2.7.2. $\mathcal{K}(X \circledast Y) = \mathcal{K}(X) * \mathcal{K}(Y)$.

Given a point x in a finite T_0-space X, the *star* C_x of x consists of the points which are comparable with x, i.e. $C_x = U_x \cup F_x$. Note that C_x is always contractible since $1_{C_x} \leq f \geq g$ where $f : C_x \to C_x$ is the map which is the identity on F_x and the constant map x on U_x, and g is the constant map x. The *link* of x is the subspace $\hat{C}_x = C_x \smallsetminus \{x\}$. In case we need to specify the ambient space X, we will write $\hat{C}_x^X$. Note that $\hat{C}_x = \hat{U}_x \circledast \hat{F}_x$.

Proposition 2.7.3. *Let X and Y be finite T_0-spaces. Then $X \circledast Y$ is contractible if and only if X or Y is contractible.*

Proof. Assume X is contractible. Then there exists a sequence of spaces

$$X = X_n \underset{\neq}{\supsetneq} X_{n-1} \underset{\neq}{\supsetneq} \ldots \underset{\neq}{\supsetneq} X_1 = \{x_1\}$$

with $X_i = \{x_1, x_2, \ldots, x_i\}$ and such that x_i is a beat point of X_i for every $2 \leq i \leq n$. Then x_i is a beat point of $X_i \circledast Y$ for each $2 \leq i \leq n$ and therefore, $X \circledast Y$ deformation retracts to $\{x_1\} \circledast Y$ which is contractible. Analogously, if Y is contractible, so is $X \circledast Y$.

Now suppose $X \circledast Y$ is contractible. Then there exists a sequence

$$X \circledast Y = X_n \circledast Y_n \underset{\neq}{\supsetneq} X_{n-1} \circledast Y_{n-1} \underset{\neq}{\supsetneq} \ldots \underset{\neq}{\supsetneq} X_1 \circledast Y_1 = \{z_1\}$$

with $X_i \subseteq X$, $Y_i \subseteq Y$, $X_i \circledast Y_i = \{z_1, z_2 \ldots, z_i\}$ such that z_i is a beat point of $X_i \circledast Y_i$ for $i \geq 2$.

Let $i \geq 2$. If $z_i \in X_i$, z_i is a beat point of X_i unless it is a maximal point of X_i and Y_i has a minimum. In the same way, if $z_i \in Y_i$, z_i is a beat point of Y_i or X_i has a maximum. Therefore, for each $2 \leq i \leq n$, either $X_{i-1} \subseteq X_i$ and $Y_{i-1} \subseteq Y_i$ are deformation retracts (in fact, one inclusion is an identity and the other inclusion is strict), or one of them, X_i or Y_i, is contractible. This proves that X or Y is contractible. $\qquad\square$

In Proposition 4.3.4 we will prove a result which is the analogue of Proposition 2.7.3 for collapsible finite spaces.

If X and Y are finite spaces, the preorder corresponding to the topological product $X \times Y$ is the product of the preorders of X and Y (Remark 1.1.2), i.e. $(x, y) \leq (x', y')$ if and only if $x \leq x'$ and $y \leq y'$. If X and Y are two topological spaces, not necessarily finite, and A is strong deformation retract of a X, then $A \times Y$ is a strong deformation retract of $X \times Y$.

Proposition 2.7.4. *Let X_c and Y_c be cores of finite spaces X and Y. Then $X_c \times Y_c$ is a core of $X \times Y$.*

Proof. Since $X_c \subseteq X$ is a strong deformation retract, so is $X_c \times Y \subseteq X \times Y$. Analogously $X_c \times Y_c$ is a strong deformation retract of $X_c \times Y$ and then, so is $X_c \times Y_c \subseteq X \times Y$. We have to prove that the product of minimal finite spaces is also minimal. Let $(x, y) \in X_c \times Y_c$. If there exists $x' \in X_c$ with $x' \prec x$ and $y' \in Y_c$ with $y' \prec y$, (x, y) covers at least two elements (x', y) and (x, y'). If x is minimal in X_c, $\hat{U}_{(x,y)}$ is homeomorphic to $\hat{U}_y$. Analogously if y is minimal. Therefore, (x, y) is not a down beat point. Similarly, $X_c \times Y_c$ does not have up beat points. Thus, it is a minimal finite space. $\qquad\square$

In particular $X \times Y$ is contractible if and only if each space X and Y is contractible. In fact this result holds in general, when X and Y are not necessarily finite.

Recall that the product of two nonempty spaces is T_0 if and only if each space is.

Proposition 2.7.5. *Let X and Y be finite T_0-spaces. Then $|\mathcal{K}(X \times Y)|$ is homeomorphic to $|\mathcal{K}(X)| \times |\mathcal{K}(Y)|$.*

Proof. Let $p_X : X \times Y \to X$ and $p_Y : X \times Y \to Y$ be the canonical projections. Define $f : |\mathcal{K}(X \times Y)| \to |\mathcal{K}(X)| \times |\mathcal{K}(Y)|$ by $f = |\mathcal{K}(p_X)| \times |\mathcal{K}(p_Y)|$. In other words, if $\alpha = \sum_{i=0}^{k} t_i(x_i, y_i) \in |\mathcal{K}(X \times Y)|$ where $(x_0, y_0) < (x_1, y_1) < \ldots < (x_k, y_k)$ is a chain in $X \times Y$, $f(\alpha) = (\sum_{i=0}^{k} t_i x_i, \sum_{i=0}^{k} t_i y_i)$.

Since $|\mathcal{K}(p_X)|$ and $|\mathcal{K}(p_Y)|$ are continuous, so is f. $|\mathcal{K}(X \times Y)|$ is compact and $|\mathcal{K}(X)| \times |\mathcal{K}(Y)|$ is Hausdorff, so we only need to show that f is a bijection. Details will be left to the reader. An explicit formula for $g = f^{-1}$ is given by

$$g\left(\sum_{i=0}^{k} u_i x_i, \sum_{i=0}^{l} v_i y_i\right) = \sum_{i,j} t_{ij}(x_i, y_j),$$

where $t_{ij} = \max\{0, \min\{u_0 + u_1 + \ldots + u_i, v_0 + v_1 + \ldots v_j\} - \max\{u_0 + u_1 + \ldots + u_{i-1}, v_0 + v_1 + \ldots v_{j-1}\}\}$. The idea is very simple. Consider the segments $U_0, U_1, \ldots, U_k \subseteq I = [0,1]$, each U_i of length u_i, $U_i = [u_0 + u_1 + \ldots + u_{i-1}, u_0 + u_1 + \ldots + u_i]$. Analogously, define $V_j = [v_0 + v_1 + \ldots + v_{j-1}, v_0 + v_1 + \ldots + v_j] \subseteq I$ for $0 \leq j \leq l$. Then t_{ij} is the length of the segment $U_i \cap V_j$. It is not hard to see that $g : |\mathcal{K}(X)| \times |\mathcal{K}(Y)| \to |\mathcal{K}(X \times Y)|$ is well defined since $support(\sum_{i,j} t_{ij}(x_i, y_j))$ is a chain and $\sum_{i,j} t_{ij} = \sum_{i,j} length(U_i \cap V_j) = \sum_i length(U_i) = 1$. Moreover, the compositions gf and fg are the corresponding identities. $\square$

A similar proof of the last result can be found in [81, Proposition 4.1].

If X is a finite T_0-space, and $A \subseteq X$ is a subspace, the quotient X/A need not be T_0. For example, if X is the chain of three elements $0 < 1 < 2$ and $A = \{0, 2\}$, X/A is the indiscrete space of two elements. We will exhibit a necessary and sufficient condition for X/A to be T_0.

Let X be a finite space and $A \subseteq X$ a subspace. We will denote by $q : X \to X/A$ the quotient map and by qx the class in the quotient of an element $x \in X$. Recall that $\overline{A} = \{x \in X \mid \exists\, a \in A \text{ with } x \geq a\}$ denotes the closure of A. We will denote by $\underline{A} = \{x \in X \mid \exists\, a \in A \text{ with } x \leq a\} = \bigcup_{a \in A} U_a \subseteq X$, the open hull of A.

Lemma 2.7.6. *Let $x \in X$. If $x \in \overline{A}$, $U_{qx} = q(U_x \cup \underline{A})$. If $x \notin \overline{A}$, $U_{qx} = q(U_x)$.*

Proof. Suppose $x \in \overline{A}$. A subset U of X/A is open if and only if $q^{-1}(U)$ is open in X. Since $q^{-1}(q(U_x \cup \underline{A})) = U_x \cup \underline{A} \subseteq X$ is open, $q(U_x \cup \underline{A}) \subseteq X/A$ is an open set containing qx. Therefore $U_{qx} \subseteq q(U_x \cup \underline{A})$. The other inclusion follows from the continuity of q since $x \in \overline{A}$: if $y \in \underline{A}$, there exist $a, b \in A$ such that $y \leq a$ and $b \leq x$ and therefore $qy \leq qa = qb \leq qx$.

If $x \notin \overline{A}$, $q^{-1}(q(U_x)) = U_x$, so $q(U_x)$ is open and therefore $U_{qx} \subseteq q(U_x)$. The other inclusion is trivial. $\square$

Proposition 2.7.7. *Let X be a finite space and $A \subseteq X$ a subspace. Let $x, y \in X$, then $qx \leq qy$ in the quotient X/A if and only if $x \leq y$ or there exist $a, b \in A$ such that $x \leq a$ and $b \leq y$.*

Proof. Assume $qx \leq qy$. If $y \in \overline{A}$, there exists $b \in A$ with $b \leq y$ and by the previous lemma $qx \in U_{qy} = q(U_y \cup \underline{A})$. Therefore $x \in U_y \cup \underline{A}$ and then $x \leq y$ or $x \leq a$ for some $a \in A$. If $y \notin \overline{A}$, $qx \in U_{qy} = q(U_y)$. Hence, $x \in U_y$.

Conversely if $x \leq y$ or there are some $a, b \in A$ such that $x \leq a$ and $b \leq y$, then $qx \leq qy$ or $qx \leq qa = qb \leq qy$. $\square$

Proposition 2.7.8. *Let X be a finite T_0-space and $A \subseteq X$. The quotient X/A is not T_0 if and only if there exists a triple $a < x < b$ with $a, b \in A$ and $x \notin A$.*

Proof. Suppose there is not such triple and that $qx \leq qy$, $qy \leq qx$. Then $x \leq y$ or there exist $a, b \in A$ with $x \leq a$, $b \leq y$, and, on the other hand, $y \leq x$ or there are some $a', b' \in A$ such that $y \leq a'$, $b' \leq x$. If $x \leq y$ and $y \leq x$, then $x = y$. In other case, both x and y are in A. Therefore, $qx = qy$. This proves that X/A is T_0. Conversely, if there exists a triple $a < x < b$ as above, $qa \leq qx \leq qb = qa$, but $qa \neq qx$. Therefore, X/A is not T_0.   □

The non-existence of a triple as above is equivalent to saying that $A = \overline{A} \cap \underline{A}$, i.e.
$$X/A \text{ is } T_0 \text{ if and only if } A = \overline{A} \cap \underline{A}.$$
For example open or closed subsets satisfy this condition.

Now we want to study how the functors $\mathcal{X}$ and $\mathcal{K}$ behave with respect to quotients. Recall that $\mathcal{K}(\mathcal{X}(K))$ is the barycentric subdivision K' of K. Following [80] and [35], the *barycentric subdivision* of a finite T_0-space X is defined by $X' = \mathcal{X}(\mathcal{K}(X))$. Explicitly, X' consists of the nonempty chains of X ordered by inclusion. This notion will be important in the development of the simple homotopy theory for finite spaces studied in Chap. 4.

Example 2.7.9. Let $X = \mathbb{C}D_2 = \{x, a, b\}$ and let $A = \{a, b\}$ be the subspace of minimal elements.

Then X/A is the Sierpinski space $\mathfrak{S}$ (the finite T_0-space with two points $0 < 1$) and $|\mathcal{K}(X)|/|\mathcal{K}(A)|$ is homeomorphic to S^1. Therefore $|\mathcal{K}(X)|/|\mathcal{K}(A)|$ and $|\mathcal{K}(X/A)|$ are not homotopy equivalent. However $X'/A' = S^0 \circledast S^0$ and then $|\mathcal{K}(X')|/|\mathcal{K}(A')|$ and $|\mathcal{K}(X'/A')|$ are both homeomorphic to a circle. The application $\mathcal{K}$ does not preserve quotients in general. In Corollary 7.2.2 we prove that if A is a subspace of a finite T_0-space X, $|\mathcal{K}(X')|/|\mathcal{K}(A')|$ and $|\mathcal{K}(X'/A')|$ are homotopy equivalent.

A particular case of a quotient X/A is the one-point union or wedge. If X and Y are topological spaces with base points $x_0 \in X$, $y_0 \in Y$, then the wedge $X \vee Y$ is the quotient $X \sqcup Y/A$ with $A = \{x_0, y_0\}$. Clearly, if X and Y are finite T_0-spaces, $A = \{x_0, y_0\} \subseteq X \sqcup Y$ satisfies $A = \overline{A} \cap \underline{A}$ and then $X \vee Y$ is also T_0. Moreover, if $x, x' \in X$, then x covers x' in X if and only if x covers x' in $X \vee Y$. The same holds for Y, and if $x \in X \smallsetminus \{x_0\}$, $y \in Y \smallsetminus \{y_0\}$ then x does not cover y in $X \vee Y$ and y does not cover x. Thus, the Hasse diagram of $X \vee Y$ is the union of the Hasse diagrams of X and Y, identifying x_0 and y_0.

If $X \vee Y$ is contractible, then X and Y are contractible. This holds for general topological spaces. Let $i : X \to X \vee Y$ denote the canonical inclusion and $r : X \vee Y \to X$ the retraction which sends all of Y to x_0. If H :

$(X \vee Y) \times I \rightarrow X \vee Y$ is a homotopy between the identity and a constant, then $rH(i \times 1_I) : X \times I \rightarrow X$ shows that X is contractible. The following example shows that the converse is not true for finite spaces.

Example 2.7.10. The space X of Example 2.2.6 is contractible, but the union at x of two copies of X is a minimal finite space, and in particular it is not contractible.

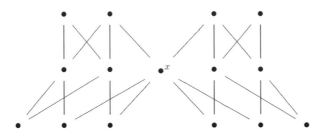

However, from Corollary 4.3.11 we will deduce that $X \vee X$ is homotopically trivial, or in other words, it is weak homotopy equivalent to a point. This is the first example we exhibit of a finite space which is homotopically trivial but which is not contractible. These spaces play a fundamental role in the theory of finite spaces.

In Proposition 4.3.10 we will prove that if X and Y are finite T_0-spaces, there is a weak homotopy equivalence $|\mathcal{K}(X)| \vee |\mathcal{K}(Y)| \rightarrow X \vee Y$.

2.8   A Finite Analogue of the Mapping Cylinder

The mapping cylinder of a map $f : X \rightarrow Y$ between topological spaces is the space Z_f obtained from $(X \times I) \sqcup Y$ by identifying each point $(x, 1) \in X \times I$ with $f(x) \in Y$. Both X and Y are subspaces of Z_f. We denote by $j : Y \hookrightarrow Z_f$ and $i : X \hookrightarrow Z_f$ the canonical inclusions where i is defined by $i(x) = (x, 0)$. The space Y is in fact a strong deformation retract of Z_f. Moreover, there exists a retraction $r : Z_f \rightarrow Y$ with $jr \simeq 1_{Z_f}$ rel Z_f which satisfies that $ri = f$ [75, Theorem 1.4.12].

We introduce a finite analogue of the classical mapping cylinder which will become important in Chap. 4. This construction was first studied in [8].

Definition 2.8.1. Let $f : X \rightarrow Y$ be a map between finite T_0-spaces. We define the *non-Hausdorff mapping cylinder* $B(f)$ as the following finite T_0-space. The underlying set is the disjoint union $X \sqcup Y$. We keep the given ordering within X and Y and for $x \in X$, $y \in Y$ we set $x \leq y$ in $B(f)$ if $f(x) \leq y$ in Y.

It can be proved that $B(f)$ is isomorphic to $(X \times \mathfrak{S}) \sqcup Y/_{(x,1)\sim f(x)}$ where $\mathfrak{S}$ denotes the Sierpinski space. However, we will omit the proof because this fact will not be used in the applications.

We will denote by $i : X \hookrightarrow B(f)$ and $j : Y \hookrightarrow B(f)$ the canonical inclusions of X and Y into the non-Hausdorff mapping cylinder.

Lemma 2.8.2. *Let $f : X \to Y$ be a map between finite T_0-spaces. Then Y is a strong deformation retract of $B(f)$.*

Proof. Define the retraction $r : B(f) \to Y$ of j by $r(x) = f(x)$ for every $x \in X$. Clearly r is order preserving. Moreover, $jr \geq 1_{B(f)}$ and then $jr \simeq 1_{B(f)}$ rel Y. $\qquad\square$

By Corollary 2.2.5, for any map $f : X \to Y$ there is a strong collapse $B(f) \searrow Y$.

Since $ri = f$, any map between finite T_0-spaces can be factorized as a composition of an inclusion and a homotopy equivalence.

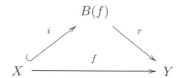

As in the classical setting, the non-Hausdorff mapping cylinder can be used to reduce many proofs concerning general maps to the case of inclusions. For example, f satisfies one of the following properties if and only if the inclusion i does: being a homotopy equivalence, a weak homotopy equivalence or a nullhomotopic map.

If X and Y are any two homotopy equivalent spaces there exists a third space Z containing both X and Y as strong deformation retracts. This space can be taken as the mapping cylinder of any homotopy equivalence $X \to Y$ (see [38, Corollary 0.21]). If $f : X \to Y$ is now a homotopy equivalence between finite T_0-spaces, Y is a strong deformation retract of $B(f)$ but X in general is just a (weak) deformation retract. Consider the space X and the point $x \in X$ of Example 2.2.6. The map $f : * \to X^{op}$ that maps $*$ into x is a homotopy equivalence. However $*$ is not a strong deformation retract of $B(f)$ by Corollary 2.2.5 because $(B(f), *)$ is a minimal pair. Although X is not in general a strong deformation retract of $B(f)$ for a homotopy equivalence $f : X \to Y$, we will see that if two finite T_0-spaces are homotopy equivalent, there exists a third finite T_0-space containing both as strong deformation retracts. This is stated in Proposition 4.6.6.

Chapter 3
Minimal Finite Models

In Sect. 2.3 we proved that in general, if K is a finite simplicial complex, there is no finite space with the homotopy type of $|K|$. However, by Theorem 1.4.12 any compact polyhedron is weak homotopy equivalent to a finite space. In this chapter we will study finite models of polyhedra in this sense and we will describe the *minimal finite models* of some well-known (Hausdorff) spaces, i.e. weak homotopy equivalent finite spaces of minimum cardinality. The main results of this chapter appear in [7].

3.1   A Finite Space Approximation

Definition 3.1.1. Let X be a space. We say that a finite space Y is a *finite model* of X if it is weak homotopy equivalent to X. We say that Y is a *minimal finite model* if it is a finite model of minimum cardinality.

For example, the singleton is the unique minimal finite model of every contractible space. Moreover, it is the unique minimal finite model of every homotopically trivial space, i.e. with trivial homotopy groups.

Since every finite space is homotopy equivalent to its core, which is a smaller space, we have the following

Remark 3.1.2. Every minimal finite model is a minimal finite space.

Since $\mathcal{K}(X) = \mathcal{K}(X^{op})$, if X is a minimal finite model of a space Y, then so is X^{op}.

Example 3.1.3. The 5-point T_0-space X, whose Hasse diagram is

J.A. Barmak, *Algebraic Topology of Finite Topological Spaces and Applications*, Lecture Notes in Mathematics 2032, DOI 10.1007/978-3-642-22003-6_3, © Springer-Verlag Berlin Heidelberg 2011

has an associated polyhedron $|\mathcal{K}(X)|$, which is homotopy equivalent to $S^1 \vee S^1$. Therefore, X is a finite model of $S^1 \vee S^1$. In fact, it is a minimal finite model since every space with less than 5 points is either contractible, or non connected or weak homotopy equivalent to S^1. However, this minimal finite model is not unique since X^{op} is another minimal finite model not homeomorphic to X.

We will generalize this result later, when we characterize the minimal finite models of graphs.

The idea of modeling spaces with ones which are easier to describe and to work with is standard in Algebraic Topology. For instance, any topological space X can be approximated by a CW-complex, in the sense that there exists a CW-complex weak homotopy equivalent to X. Moreover, two such CW-approximations are homotopy equivalent [38, Proposition 4.13, Corollary 4.19]. In these notes, we use finite spaces to model general spaces. Note that, as the previous example shows, two finite models of a space may not be homotopy equivalent. The uniqueness of CW-approximations up to homotopy implies that if X is a finite model of a polyhedron Y, then Y is homotopy equivalent to $|\mathcal{K}(X)|$.

Generalizing the definition made in Sect. 2.7, we define the *non-Hausdorff suspension* $\mathbb{S}(X)$ of a topological space X as the space $X \cup \{+, -\}$ whose open sets are those of X together with $X \cup \{+\}$, $X \cup \{-\}$ and $X \cup \{+, -\}$. If X is a finite space, the non-Hausdorff suspension of X is the join $\mathbb{S}(X) = X \circledast S^0$. The *non-Hausdorff suspension of order* n is defined recursively by $\mathbb{S}^n(X) = \mathbb{S}(\mathbb{S}^{n-1}(X))$. For convenience we define $\mathbb{S}^0(X) = X$.

The following result is due to McCord [55].

Proposition 3.1.4. *The finite space $\mathbb{S}^n(S^0)$ is a finite model of the n-dimensional sphere S^n for every $n \geq 0$.*

Proof. By Remark 2.7.2, $|\mathcal{K}(\mathbb{S}^n(S^0))| = |\mathcal{K}(S^0 \circledast S^0 \circledast \ldots \circledast S^0)| = |\mathcal{K}(S^0)| * |\mathcal{K}(S^0)| * \ldots * |\mathcal{K}(S^0)| = S^0 * S^0 * \ldots * S^0 = S^n$. $\qquad\qquad\square$

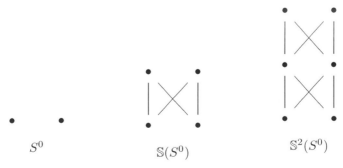

S^0 $\qquad\qquad\qquad\qquad$ $\mathbb{S}(S^0)$ $\qquad\qquad\qquad$ $\mathbb{S}^2(S^0)$

In [52] May conjectured that $\mathbb{S}^n(S^0)$ is a minimal finite model of S^n. We will show that this conjecture is true. In fact, we prove a stronger result. Namely, we will see that any space with the same homotopy groups as S^n

has at least $2n + 2$ points. Moreover, if it has exactly $2n + 2$ points then it has to be homeomorphic to $\mathbb{S}^n S^0$.

3.2 Minimal Finite Models of the Spheres

Recall again that the height $ht(X)$ of a finite poset X is one less than the maximum cardinality of a chain of X. Therefore $ht(X)$ coincides with the dimension of the associated complex $\mathcal{K}(X)$.

Theorem 3.2.1. *Let $X \neq *$ be a minimal finite space. Then X has at least $2ht(X) + 2$ points. Moreover, if X has exactly $2ht(X) + 2$ points, then it is homeomorphic to $\mathbb{S}^{ht(X)}(S^0)$.*

Proof. Let $x_0 < x_1 < \ldots < x_h$ be a chain in X of length $h = ht(X)$. Since X is a minimal finite space, x_i is not an up beat point for any $0 \leq i < h$. Then, for every $0 \leq i < h$ there exists $y_{i+1} \in X$ such that $y_{i+1} > x_i$ and $y_{i+1} \not\geq x_{i+1}$. We assert that the points y_i (for $0 < i \leq h$) are all distinct from each other and also different from the x_j ($0 \leq j \leq h$).

Since $y_{i+1} > x_i$, it follows that $y_{i+1} \neq x_j$ for all $j \leq i$. But $y_{i+1} \neq x_j$ for all $j > i$ because $y_{i+1} \not\geq x_{i+1}$.

If $y_{i+1} = y_{j+1}$ for some $i < j$, then $y_{i+1} = y_{j+1} \geq x_j \geq x_{i+1}$, which is a contradiction.

Since finite spaces with minimum or maximum are contractible and $X \neq *$ is a minimal finite space, it cannot have a minimum. Then there exists $y_0 \in X$ such that $y_0 \not\geq x_0$. Therefore, y_0 must be distinct from the other $2h+1$ points and $\#X \geq 2h + 2$.

Let us suppose now that X has exactly $2h + 2$ points, i.e.

$$X = \{x_0, x_1, \ldots, x_h, y_0, y_1, \ldots, y_h\}.$$

Because of the maximality of the chain $x_0 < \ldots < x_h$, we get that x_i and y_i are incomparable for all i.

We show that $y_i < x_j$ and $y_i < y_j$ for all $i < j$ by induction in j.

For $j = 0$ there is nothing to prove. Let $0 \leq k < h$ and assume the statement holds for $j = k$. As x_{k+1} is not a down beat point, there exists $z \in X$ such that $z < x_{k+1}$, and $z \not\leq x_k$. Since x_{k+1} and y_{k+1} are incomparable, it follows that $z \neq y_{k+1}$. By induction we know that every point in X, with the exception of y_k and y_{k+1}, is greater than x_{k+1} or less than x_k. Then $z = y_k$ and so, $y_k < x_{k+1}$. Analogously, y_{k+1} is not a down beat point and there exists $w \in X$ such that $w < y_{k+1}$ and $w \not\leq x_k$. Again by induction, and because $y_{k+1} \not\geq x_{k+1}$, we deduce that w must be y_k and then $y_k < y_{k+1}$. Furthermore, if $i < k$, then $y_i < x_k < x_{k+1}$ and $y_i < x_k < y_{k+1}$.

We proved that, for any $i < j$, we have that $y_i < x_j$, $y_i < y_j$, $x_i < x_j$ and $x_i < y_j$. Moreover, for any $0 \leq i \leq h$, x_i and y_i are incomparable.

This is exactly the order of $\mathbb{S}^h(S^0)$. Therefore X is homeomorphic to $\mathbb{S}^h(S^0)$. $\qquad\square$

Theorem 3.2.2. *Any space with the same homotopy groups as S^n has at least $2n + 2$ points. Moreover, $\mathbb{S}^n(S^0)$ is the unique space with $2n + 2$ points with this property.*

Proof. The case $n = 1$ is trivial. In the other cases, let us suppose that X is a finite space with minimum cardinality such that $\pi_k(X, x) = \pi_k(S^n, s)$ for all $k \geq 0$. Then X must be a minimal finite space and so is T_0.

By the Hurewicz Theorem [38, Theorem 4.32], $H_n(|\mathcal{K}(X)|) = \pi_n(|\mathcal{K}(X)|) = \pi_n(S^n) \neq 0$. This implies that the dimension of the simplicial complex $\mathcal{K}(X)$ must be at least n, which means that the height of X is at least n. The result now follows immediately from the previous theorem. $\qquad\square$

Corollary 3.2.3. *The n-sphere has a unique minimal finite model and it has $2n + 2$ points.*

Remark 3.2.4. These results regarding the minimal finite models of the spheres were obtained in [7]. However, there is an article of McCord [54] with a result without proof [54, Theorem 2], from which the first part of Theorem 3.2.2 could be deduced. McCord's result can be easily deduced from the stronger Theorem 3.2.1 (which also implies the uniqueness of these minimal models).

Furthermore, the proof of Theorem 3.2.1 itself is interesting because it relates the combinatorial methods of Stong's theory with McCord's point of view.

3.3 Minimal Finite Models of Graphs

Remark 3.3.1. If X is a connected finite T_0-space of height one, $|\mathcal{K}(X)|$ is a connected graph, i.e. a CW-complex of dimension one. Therefore, the weak homotopy type of X is completely determined by its Euler characteristic. More precisely, if

$$\chi(X) = \#X - \#\mathrm{E}(\mathcal{H}(X)) = n,$$

then X is a finite model of $\bigvee_{i=1}^{1-n} S^1$. Recall that $\mathrm{E}(\mathcal{H}(X))$ denotes the set of edges of the Hasse diagram of X.

Proposition 3.3.2. *Let X be a connected finite T_0-space and let $x_0, x \in X$, $x_0 \neq x$ such that x is neither maximal nor minimal in X. Then the inclusion map of the associated simplicial complexes $\mathcal{K}(X \smallsetminus \{x\}) \subseteq \mathcal{K}(X)$ induces an epimorphism*

$$i_* : E(\mathcal{K}(X \smallsetminus \{x\}), x_0) \to E(\mathcal{K}(X), x_0)$$

between their edge-path groups.

Proof. We have to check that every closed edge-path in $\mathcal{K}(X)$ with base point x_0 is equivalent to another edge-path that does not go through x. Let us suppose that $y \le x$ and $(y, x)(x, z)$ is an edge-path in $\mathcal{K}(X)$. If $x \le z$ then $(y, x)(x, z) \equiv (y, z)$. In the case that $z < x$, since x is not maximal in X, there exists $w > x$. Therefore $(y, x)(x, z) \equiv (y, x)(x, w)(w, x)(x, z) \equiv (y, w)(w, z)$. The case $y \ge x$ is analogous.

In this way, one can eliminate x from the writing of any closed edge-path with base point x_0.   □

Note that the space $X \smallsetminus \{x\}$ of the previous proposition is also connected. An alternative proof of the previous proposition is given by the van Kampen Theorem. Let $C_x = U_x \cup F_x$ be the star of x. Since x is not maximal or minimal, the link $\hat{C}_x = C_x \smallsetminus \{x\}$ is connected. Then van Kampen gives an epimorphism $\pi_1(|\mathcal{K}(X \smallsetminus x)|) * \pi_1(|\mathcal{K}(C_x)|) \to \pi_1(|\mathcal{K}(X)|)$. But $\mathcal{K}(C_x) = x\mathcal{K}(\hat{C}_x)$ is a cone, and then $\pi_1(|\mathcal{K}(C_x)|) = 0$. Therefore, $i_* : \pi_1(|\mathcal{K}(X \smallsetminus x)|) \to \pi_1(|\mathcal{K}(X)|)$ is an epimorphism.

The result above shows one of the advantages of using finite spaces instead of simplicial complexes. The conditions of maximality or minimality of points in a finite space are hard to express in terms of simplicial complexes.

Remark 3.3.3. If X is a finite T_0-space, then $ht(X) \le 1$ if and only if every point in X is maximal or minimal.

Corollary 3.3.4. *Let X be a connected finite space. Then there exists a connected T_0-subspace $Y \subseteq X$ of height at most one such that the fundamental group of X is a quotient of the fundamental group of Y.*

Proof. We can assume that X is T_0 because X has a core. Now, the result follows immediately from the previous proposition.   □

Remark 3.3.5. Note that the fundamental group of a connected finite T_0-space of height at most one is finitely generated by Remark 3.3.1. Therefore, path-connected spaces whose fundamental group does not have a finite set of generators do not admit finite models.

Corollary 3.3.6. *Let $n \in \mathbb{N}$. If X is a minimal finite model of $\bigvee_{i=1}^{n} S^1$, then $ht(X) = 1$.*

Proof. Let X be a minimal finite model of $\bigvee_{i=1}^{n} S^1$. Then there exists a connected T_0-subspace $Y \subseteq X$ of height one, $x \in Y$ and an epimorphism from $\pi_1(Y, x)$ to $\pi_1(X, x) = \underset{i=1}{\overset{n}{*}} \mathbb{Z}$.

Since $ht(Y) = 1$, Y is a model of a graph, thus $\pi_1(Y, x) = \overset{m}{\underset{i=1}{*}} \mathbb{Z}$ for some integer m. Note that $m \geq n$.

There are m edges of $\mathcal{H}(Y)$ which are not in a maximal tree of the underlying non directed graph of $\mathcal{H}(Y)$ (i.e. $\mathcal{K}(Y)$). Therefore, we can remove $m - n$ edges from $\mathcal{H}(Y)$ in such a way that it remains connected and the new space Z obtained in this way is a model of $\overset{n}{\underset{i=1}{\bigvee}} S^1$.

Note that $\#Z = \#Y \leq \#X$, but since X is a minimal finite model, $\#X \leq \#Z$ and then $X = Y$ has height one. $\qquad\qquad\qquad\qquad\square$

If X is a minimal finite model of $\overset{n}{\underset{i=1}{\bigvee}} S^1$ and we call $i = \#\{y \in X \mid y$ is maximal$\}$, $j = \#\{y \in X \mid y$ is minimal$\}$, then $\#X = i+j$ and $\#\mathrm{E}(\mathcal{H}(X)) \leq ij$. Since $\chi(X) = 1 - n$, we have that $n \leq ij - (i+j) + 1 = (i-1)(j-1)$.

We can now state the main result of this section.

Theorem 3.3.7. *Let $n \in \mathbb{N}$. A finite T_0-space X is a minimal finite model of $\overset{n}{\underset{i=1}{\bigvee}} S^1$ if and only if $ht(X) = 1$, $\#X = min\{i+j \mid (i-1)(j-1) \geq n\}$ and $\#\mathrm{E}(\mathcal{H}(X)) = \#X + n - 1$.*

Proof. We have already proved that if X is a minimal finite model of $\overset{n}{\underset{i=1}{\bigvee}} S^1$, then $ht(X) = 1$ and $\#X \geq min\{i+j \mid (i-1)(j-1) \geq n\}$. If i and j are such that $n \leq (i-1)(j-1)$, we can consider $Y = \{x_1, x_2, \ldots, x_i, y_1, y_2, \ldots y_j\}$ with the order $y_k \leq x_l$ for all k, l, which is a model of $\overset{(i-1)(j-1)}{\underset{k=1}{\bigvee}} S^1$. Then we can remove $(i-1)(j-1) - n$ edges from $\mathcal{H}(X)$ to obtain a connected space of cardinality $i+j$ which is a finite model of $\overset{n}{\underset{k=1}{\bigvee}} S^1$. Therefore $\#X \leq \#Y = i+j$. This is true for any i, j with $n \leq (i-1)(j-1)$, then $\#X = min\{i+j \mid (i-1)(j-1) \geq n\}$. Moreover, $\#\mathrm{E}(\mathcal{H}(X)) = \#X + n - 1$ because $\chi(X) = 1 - n$.

In order to show the converse of the theorem we only need to prove that the conditions $ht(X) = 1$, $\#X = min\{i + j \mid (i - 1)(j - 1) \geq n\}$ and $\#\mathrm{E}(\mathcal{H}(X)) = \#X + n - 1$ imply that X is connected, because in this case, by Remark 3.3.1, the first and third conditions would say that X is a model of $\overset{n}{\underset{i=1}{\bigvee}} S^1$, and the second condition would say that it has the right cardinality.

Suppose X satisfies the conditions of above and let X_l, $1 \leq l \leq k$, be the connected components of X. Let us denote by M_l the set of maximal elements of X_l and let $m_l = X_l \smallsetminus M_l$. Let $i = \sum_{r=1}^{k} \#M_l$, $j = \sum_{r=1}^{k} \#m_l$. Since $i+j = \#X = min\{s+t \mid (s-1)(t-1) \geq n\}$, it follows that $(i-2)(j-1) < n = \#\mathrm{E}(\mathcal{H}(X)) - \#X + 1 = \#\mathrm{E}(\mathcal{H}(X)) - (i+j) + 1$. Hence $ij - \#\mathrm{E}(\mathcal{H}(X)) < j - 1$. This means that $\mathcal{K}(X)$ differs from the complete bipartite graph $(\cup m_l, \cup M_l)$ in less than $j - 1$ edges. Since there are no edges from m_r to M_l if $r \neq l$,

$$j - 1 > \sum_{l=1}^{k} \#M_l(j - \#m_l) \geq \sum_{l=1}^{k}(j - \#m_l) = (k - 1)j.$$

Therefore $k = 1$ and the proof is complete. $\qquad\square$

For a real number r, we denote by $\lceil r \rceil = \min\{m \in \mathbb{Z} \mid m \geq r\}$ the ceiling of r.

Corollary 3.3.8. *The cardinality of a minimal finite model of $\bigvee_{i=1}^{n} S^1$ is*

$$min\{2\lceil \sqrt{n} + 1 \rceil, 2\left\lceil \frac{1 + \sqrt{1 + 4n}}{2} \right\rceil + 1\}.$$

Proof. The minimum $m = \min\{i+j \mid (i-1)(j-1) \geq n\}$ is attained for $i = j$ or $i = j+1$ since $i(j-2) > (i-1)(j-1)$ if $j \geq i+2$. Therefore m is the minimum between $\min\{2i \mid (i-1)^2 \geq n\} = 2\lceil \sqrt{n}+1 \rceil$ and $\min\{2i+1 \mid i(i-1) \geq n\} = 2\lceil \frac{1+\sqrt{1+4n}}{2} \rceil + 1$. $\qquad\square$

Note that a space may admit many minimal finite models as we can see in the following example.

Example 3.3.9. Any minimal finite model of $\bigvee_{i=1}^{3} S^1$ has 6 points and 8 edges. So, they are, up to homeomorphism

In fact, using our characterization, we prove the following

Proposition 3.3.10. $\bigvee_{k=1}^{n} S^1$ *has a unique minimal finite model if and only if n is a square.*

Proof. Assume that $n = m^2$ is a square. In this case the cardinality of a minimal finite model X is $2m + 2$ and the numbers i and j of maximal and minimal elements have to be equal to $m+1$ in order to satisfy $(i-1)(j-1) \geq m^2$. The number of edges in the Hasse diagram is $\#E(\mathcal{H}(X)) = \#X + n - 1 = 2m + 2 + m^2 - 1 = (m + 1)^2$, and therefore every maximal element in X is greater than any minimal element. Thus, X is the non-Hausdorff join of two discrete spaces of $m + 1$ points. Conversely, suppose that n is not a square and let X be a minimal finite model of $\bigvee_{k=1}^{n} S^1$. If X^{op} is not homeomorphic to X, we have found a different minimal finite model of $\bigvee_{k=1}^{n} S^1$.

Assume then that X and X^{op} are homeomorphic and, in particular, X has the same number i of maximal and minimal elements. Since n is not a square, $\#E(\mathcal{H}(X)) = \#X + n - 1 = 2i + n - 1 \neq i^2$. Then we construct a new space Y of height 1, with $i - 1$ maximal elements, $i + 1$ minimal and with exactly $\#E(\mathcal{H}(Y)) = \#E(\mathcal{H}(X))$ edges in the Hasse diagram. This can be done because $(i - 1)(i + 1) = i^2 - 1 \geq E(\mathcal{H}(X))$. By Theorem 3.3.7, Y is a minimal finite model of $\bigvee_{k=1}^{n} S^1$ which is different from X. $\square$

Recall that an Eilenberg-MacLane space $K(G, n)$ is a space with a unique non-trivial homotopy group $\pi_n(K(G, n)) = G$. The homotopy type of a CW-complex $K(G, n)$ is uniquely determined by G and n (see [38, Proposition 4.30] for example). Note that since any graph is a $K(G, 1)$, the minimal finite models of a graph X are, in fact, the smallest spaces with the same homotopy groups as X.

The 2-dimensional torus $S^1 \times S^1 = K(\mathbb{Z} \oplus \mathbb{Z}, 1)$ is another example of an Eilenberg-MacLane space. The difficulty of finding minimal finite models of spaces can be already encountered in this particular case. The product of two minimal finite models of S^1 has 16 points, it is a finite model of $S^1 \times S^1$ and it is a minimal finite space. However it is still unknown whether this is a minimal finite model or not.

3.4   The $f^{\infty}(X)$

Stong proved that any homotopy equivalence between minimal finite spaces is a homeomorphism (Corollary 1.3.7). In this section we introduce the construction $f^{\infty}(X)$ and we exhibit some of its properties. One interesting application is an analogue of Stong's result for weak homotopy equivalences and minimal finite models.

Definition 3.4.1. Let X be a finite T_0-space and $f : X \to X$ a continuous map. We define $f^{\infty}(X) = \bigcap_{i \geq 1} f^i(X) \subseteq X$.

Remark 3.4.2. Given $f : X \to X$, there exists $n_0 \in \mathbb{N}$ such that $n \geq n_0$ implies $f^n(X) = f^{\infty}(X)$. Let $k \in \mathbb{N}$ be the order of $f|_{f^{\infty}(X)}$ in the finite group $Aut(f^{\infty}(X))$. If $n \geq n_0$ and k divides n, $f^n(X) = f^{\infty}(X)$ and $f^n|_{f^{\infty}(X)} = 1_{f^{\infty}(X)}$. In this case we will say that $n \in \mathbb{N}$ is *suitable* for f.

Remark 3.4.3. $f^{\infty}(X) = \{x \in X \mid \exists\, n \in \mathbb{N}$ such that $f^n(x) = x\}$.

Proposition 3.4.4. *Let X be a finite T_0-space and let $f, g : X \to X$ be two homotopic maps. Then $f^{\infty}(X)$ is homotopy equivalent to $g^{\infty}(X)$.*

Proof. We can assume that $g \leq f$. By Remark 3.4.2, there exists $n \in \mathbb{N}$ which is suitable for f and g simultaneously. Then one can consider $f^n|_{g^\infty(X)} : g^\infty(X) \to f^\infty(X)$ and $g^n|_{f^\infty(X)} : f^\infty(X) \to g^\infty(X)$. Since

$$f^n|_{g^\infty(X)} g^n|_{f^\infty(X)} \leq f^{2n}|_{f^\infty(X)} = 1_{f^\infty(X)},$$

$f^n|_{g^\infty(X)} g^n|_{f^\infty(X)} \simeq 1_{f^\infty(X)}$. Analogously, $g^n|_{f^\infty(X)} f^n|_{g^\infty(X)} \simeq 1_{g^\infty(X)}$. $\quad\square$

Proposition 3.4.5. *Let X be a finite T_0-space and let $Y \subseteq X$ be a subspace. Then there exists a continuous map $f : X \to X$ such that $f^\infty(X) = Y$ if and only if Y is a retract of X.*

Proof. If $Y = f^\infty(X)$ for some f, choose $n \in \mathbb{N}$ suitable for f. Then $f^n : X \to Y$ is a retraction. Conversely, if $r : X \to Y$ is a retraction, $r^\infty(X) = Y$.
$\quad\square$

Example 3.4.6. Let X be the following finite T_0-space

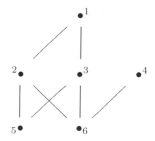

Define $f : X \to X$ such that 5 and 6 are fixed, $f(1) = f(2) = f(3) = 2$, $f(4) = 3$. Since X is contractible and $f(X)$ is a finite model of S^1, $f(X)$ is not a retract of X. However, $f^\infty(X) = \{2, 5, 6\}$ is a retract of X.

Theorem 3.4.7. *Let X be a finite T_0-space and let $f : X \to X$ be a weak homotopy equivalence. Then the inclusion $i : f^\infty(X) \hookrightarrow X$ is a weak homotopy equivalence. In particular, if X is a minimal finite model, f is a homeomorphism.*

Proof. Let $n \in \mathbb{N}$ be suitable for f. Then $f^n : X \to f^\infty(X)$, and the compositions $f^n i = 1_{f^\infty(X)}$, $if^n = f^n : X \to X$ are weak homotopy equivalences. Therefore i is a weak homotopy equivalence.

If X is a minimal finite model, $f^\infty(X) \subseteq X$ cannot have less points than X, then $f^\infty(X) = X$ and $f : X \to X$ is onto. Therefore, it is a homeomorphism.
$\quad\square$

Observe that with the same proof of the last theorem, one can prove that if $f : X \to X$ is a homotopy equivalence, then $i : f^\infty(X) \hookrightarrow X$ is a homotopy equivalence. In particular, if X is a representative of minimum cardinality of its homotopy type (ie, a minimal finite space), f is a homeomorphism.

This proves part of Stong's Classification Theorem 1.3.7 without using beat points.

Corollary 3.4.8. *Let X and Y be minimal finite models. Suppose there exist weak homotopy equivalences $f : X \to Y$ and $g : Y \to X$. Then f and g are homeomorphisms.*

Proof. The composition $gf : X \to X$ is a weak homotopy equivalence and then a homeomorphism by Theorem 3.4.7. Analogously gf is a homeomorphism. Then the result follows.   □

Open problem: Is every weak homotopy equivalence between minimal finite models a homeomorphism?

Remark 3.4.9. In Example 1.4.17 we proved that there is no weak homotopy equivalence $\mathbb{S}(D_3) \to \mathbb{S}(D_3)^{op}$. We give here an alternative proof using the previous result and our description of the minimal finite models of graphs.

Suppose there exists a weak homotopy equivalence $f : \mathbb{S}(D_3) \to \mathbb{S}(D_3)^{op}$. Then f^{op} is also a weak homotopy equivalence. Since $\mathbb{S}(D_3)$ is a minimal finite model (see Sect. 3.3), so is $\mathbb{S}(D_3)^{op}$. By Corollary 3.4.8, $\mathbb{S}(D_3)$ is homeomorphic to its opposite, which is clearly absurd.

Proposition 3.4.10. *Let X be a finite T_0-space and $f, g : X \to X$ two maps. Then $(gf)^{\infty}(X)$ and $(fg)^{\infty}(X)$ are homeomorphic.*

Proof. Let $x \in (gf)^{\infty}(X)$, then there exists $n \in \mathbb{N}$ such that $(gf)^n(x) = x$. Therefore $(fg)^n(f(x)) = f(x)$, and $f(x) \in (fg)^{\infty}(X)$. Then $f|_{(gf)^{\infty}(X)} : (gf)^{\infty}(X) \to (fg)^{\infty}(X)$. Analogously $g|_{(fg)^{\infty}(X)} : (fg)^{\infty}(X) \to (gf)^{\infty}(X)$. The compositions of these two maps are homeomorphisms, and therefore, they are also homeomorphisms.   □

Remark 3.4.11. Let X be a finite T_0-space, and $f : X \to X$ a map. Then $(f')^{\infty}(X') = f^{\infty}(X)'$. Here $f' : X' \to X'$ denotes the map $\mathcal{X}(\mathcal{K}(f))$. A chain $x_1 < x_2 < \ldots < x_k$ is in $(f')^{\infty}(X')$ if and only if there exists n such that $(f')^n(\{x_1, x_2, \ldots, x_k\}) = \{x_1, x_2, \ldots, x_k\}$. This is equivalent to saying that there exists n such that $f^n(x_i) = x_i$ for every $1 \le i \le k$ or in other words, that $\{x_1, x_2, \ldots, x_k\} \subseteq f^{\infty}(X)$.

To finish this chapter, we introduce a nice generalization of the construction of $f^{\infty}(X)$ for the case of composable maps not necessarily equal nor with the same domain or codomain.

Suppose $X_0 \xrightarrow{f_0} X_1 \xrightarrow{f_1} \ldots$ is a sequence of maps between finite spaces. Define $Y_n = f_{n-1}f_{n-2} \cdots f_0(X_0) \subseteq X_n$ the image of the composition of the first n maps of the sequence.

Proposition 3.4.12. *There exists $n_0 \in \mathbb{N}$ such that Y_n is homeomorphic to Y_{n_0} for every $n \ge n_0$.*

Proof. Since $(\#Y_n)_{n\in\mathbb{N}}$ is a decreasing sequence, there exists $n_1 \in \mathbb{N}$ such that $\#Y_n$ is constant for $n \geq n_1$. Therefore $f_n : Y_n \to Y_{n+1}$ is a bijection for $n \geq n_1$.

Let $C_n = \{(x, x') \in Y_n \times Y_n \mid x \leq x'\}$. The map $f_n : Y_n \to Y_{n+1}$ induces a one-to-one function $F_n : C_n \to C_{n+1}$, $F_n(x, x') = (f_n(x), f_n(x'))$ for $n \geq n_1$. Therefore $(\#C_n)_{n \geq n_1}$ is increasing and bounded by $(\#Y_{n_1})^2$. Hence, there exists $n_0 \geq n_1$ such that F_n is a bijection and then $f_n : Y_n \to Y_{n+1}$ a homeomorphism for $n \geq n_0$.   □

The space Y_{n_0} constructed above is well defined up to homeomorphism and it is denoted by $(f_n)_{n\in\mathbb{N}}^\infty(X_0)$. We show that in the case that all the spaces X_n are equal, i.e. $X_n = X$ for every $n \geq 0$, $(f_n)_{n\in\mathbb{N}}^\infty(X)$ is a retract of X, as in the original case. Since X is finite, there exists a subspace $Y \subseteq X$ and an increasing sequence $(n_i)_{i\in\mathbb{N}}$ of positive integers such that $Y_{n_i} = Y$ for every $i \in \mathbb{N}$. Let $g_i = f_{n_i-1}f_{n_i-2}\ldots f_{n_1}|_{Y_{n_1}} : Y_{n_1} \to Y_{n_i}$. These maps are permutations of the finite set Y, therefore there are two equal, say $g_i = g_j$ with $i < j$. Then $f_{n_j-1}f_{n_j-2}\ldots f_{n_i}|_{Y_{n_i}} = 1_Y$, so Y is a retract of X.

Chapter 4
Simple Homotopy Types and Finite Spaces

Whitehead's theory of simple homotopy types is inspired by Tietze's theorem in combinatorial group theory, which states that any finite presentation of a group could be deformed into any other by a finite sequence of elementary moves, which are now called Tietze transformations. Whitehead translated these algebraic moves into the well-known geometric moves of elementary collapses and expansions of finite simplicial complexes. His beautiful theory turned out to be fundamental for the development of piecewise-linear topology: The s-cobordism theorem, Zeeman's conjecture [87], the applications of the theory in surgery, Milnor's classical paper on Whitehead Torsion [58] and the topological invariance of torsion represent some of its major uses and advances.

In this chapter we show how to use finite topological spaces to study simple homotopy types using the relationship between finite spaces and simplicial complexes.

We have seen that if two finite T_0-spaces X, Y are homotopy equivalent, their associated simplicial complexes $\mathcal{K}(X), \mathcal{K}(Y)$ are also homotopy equivalent. Furthermore, Osaki [65] showed that in this case, the latter have the same simple homotopy type. Nevertheless, we noticed that the converse of this result is not true in general: There are finite spaces with different homotopy types whose associated simplicial complexes have the same simple homotopy type. Starting from this point, we looked for the relation that X and Y should satisfy for their associated complexes to be simple homotopy equivalent. More specifically, we wanted to find an elementary move in the setting of finite spaces (if it existed) which corresponds exactly to a simplicial collapse of the associated polyhedra.

We discovered this elementary move when we were looking for a homotopically trivial finite space (i.e. weak homotopy equivalent to a point) which was non-contractible. In order to construct such a space, we developed a method of reduction, i.e. a method that allows us to reduce a finite space to a smaller weak homotopy equivalent space. This method of reduction together with

J.A. Barmak, *Algebraic Topology of Finite Topological Spaces and Applications*, Lecture Notes in Mathematics 2032,
DOI 10.1007/978-3-642-22003-6_4, © Springer-Verlag Berlin Heidelberg 2011

the homotopically trivial and non-contractible space (of 11 points) that we found are exhibited in Sect. 4.2. Surprisingly, this method, which consists of removing a *weak point* of the space (see Definition 4.2.2), turned out to be the key to solve the problem of translating simplicial collapses into this setting.

In Sect. 4.2 we introduce the notions of collapse and simple homotopy type in the context of finite spaces. Theorem 4.2.11 establishes the relationship between these concepts and their simplicial analogues.

We are now able to study finite spaces using all the machinery of Whitehead's simple homotopy theory for CW-complexes. But also, what is more important, we can use finite spaces to strengthen the classical theory. The elementary move in this setting is much simpler to handle and describe because it consists of adding or removing just one single point. Applications of this theorem will appear constantly in the following chapters.

In the fourth section of this chapter we investigate the class of maps between finite spaces which induce simple homotopy equivalences between their associated simplicial complexes. To this end, we introduce the notion of a *distinguished* map. Similarly to the classical case, the class of simple homotopy equivalences between finite spaces can be generated, in a certain way, by expansions and a kind of formal homotopy inverse of expansions. Remarkably this class, denoted by $\mathcal{S}$, is also generated by the distinguished maps.

Many of the results of this chapter were originally published in [8], but we exhibit here more applications and, in some cases, shorter proofs.

4.1   Whitehead's Simple Homotopy Types

At the end of the 1930s Whitehead started to investigate a combinatorial approach to homotopy theory of polyhedra introducing the notions of simplicial collapse and expansion. These moves preserve the homotopy type of the complex and therefore it is natural to ask whether any two homotopy equivalent complexes can be connected by a chain of expansions and collapses. In general, this is not true. There exists an obstruction called the *Whitehead group* which can be defined in a geometrical or in an algebraic way. If the Whitehead group $Wh(K)$ of the simplicial complex K is trivial, then any complex homotopy equivalent to K is also *simple homotopy equivalent* to K, meaning that L can be obtained from K by a sequence of expansions and collapses. A *simple homotopy equivalence* is a map which is obtained from simplicial expansions by allowing one to take compositions, homotopic maps and homotopy inverses. Given a finite simplicial complex K, every homotopy equivalence $f : |K| \to |L|$ is a simple homotopy equivalence if and only if $Wh(K) = 0$.

Although the motivating questions were raised in the setting of simplicial complexes, the development of the theory of CW-complexes allowed the simple homotopy theory to reach its maturity in Whitehead's article of 1950 [86].

In this section we will recall some basic notions on simplicial complexes and simple homotopy theory for complexes and we will fix the notations that will be used henceforth. The standard references are Whitehead's papers [84–86], Milnor's article [58] and Cohen's book [23].

If K is a simplicial complex and v is a vertex of K, the *(simplicial) star* of v in K is the subcomplex $st(v) \subseteq K$ of simplices $\sigma \in K$ such that $\sigma \cup \{v\} \in K$. The *link* of v in K is the subcomplex $lk(v) \subseteq st(v)$ of the simplices which do not contain v.

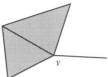

The star $st(v)$ of v

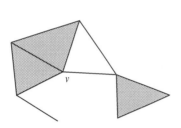

A complex K and a vertex $v \in K$

The link $lk(v)$ of v

More generally, if σ is a simplex of K, its *star* $st(\sigma)$ is the subcomplex of K whose simplices are the simplices $\tau \in K$ such that $\sigma \cup \tau \in K$. The *link* $lk(\sigma)$ is the subcomplex of $st(\sigma)$ of the simplices which are disjoint from σ.

If σ is a simplex of K, $\dot\sigma$ denotes its boundary and σ^c denotes the subcomplex of K of the simplices which do not contain σ. The *stellar subdivision* of K at the simplex σ is the complex $a\dot\sigma lk(\sigma) + \sigma^c$ where a is a vertex which is not in K. The first barycentric subdivision K' of K can be obtained from K by performing a sequence of stellar subdivisions (see [31]).

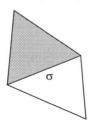

A complex K and a simplex $\sigma \in K$.

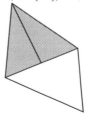

The stellar subdivision of K at σ.

Let L be a subcomplex of a finite simplicial complex K. There is an *elementary simplicial collapse* from K to L if there is a simplex σ of K and a vertex a of K not in σ such that $K = L \cup a\sigma$ and $L \cap a\sigma = a\dot{\sigma}$. This is equivalent to saying that there are only two simplices σ, τ of K which are not in L and such that τ is the unique simplex containing σ properly. In this case we say that σ is a *free face* of τ. Elementary collapses will be denoted, as usual, $K \searrow^e L$.

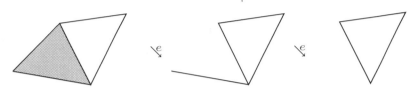

Fig. 4.1 A complex which collapses to the boundary of a 2-simplex

We say that K *(simplicially) collapses* to L (or that L *expands* to K) if there exists a sequence $K = K_1, K_2, \ldots, K_n = L$ of finite simplicial complexes such that $K_i \searrow^e K_{i+1}$ for all i (see Fig. 4.1). This is denoted by $K \searrow L$ or $L \nearrow K$. Two complexes K and L have the same *simple homotopy type* (or they are *simple homotopy equivalent*) if there is a sequence $K = K_1, K_2, \ldots, K_n = L$ such that $K_i \searrow K_{i+1}$ or $K_i \nearrow K_{i+1}$ for all i. Following Cohen's notation, we denote this by $K \diagdown\!\!\diagup L$.

Note that if there is an elementary collapse $K \searrow^e L$ where $K = L \cup a\sigma$ and $L \cap a\sigma = a\dot{\sigma}$, then the inclusion $L \cap a\sigma \hookrightarrow a\sigma$ is trivially a homotopy equivalence. By the gluing theorem A.2.5 the inclusion $L \hookrightarrow K$ is a homotopy equivalence. Therefore simple homotopy equivalent complexes are in particular homotopy equivalent.

A simplicial complex is *collapsible* if it collapses to one of its vertices. For instance, any simplicial cone aK is collapsible. If σ is a maximal simplex of K, then σ is a free face of $a\sigma$ in aK. Therefore, $aK \searrow aK \smallsetminus \{\sigma, a\sigma\} = a(K \smallsetminus \{\sigma\})$. By induction $a(K \smallsetminus \{\sigma\})$ is collapsible and then so is aK.

Lemma 4.1.1. *Let aK be a simplicial cone of a finite complex K. Then, K is collapsible if and only if $aK \searrow K$.*

Proof. We can prove by induction that for any subcomplex $L \subseteq K$, $K \searrow L$ if and only if $aK \searrow aL \cup K$. Note that σ is a free face of τ in L if and only if $a\sigma$ is a free face of $a\tau$ in $aL \cup K$. In particular K collapses to a vertex v if and only if $aK \searrow av \cup K$. Since a is a free face of av in $av \cup K$, $av \cup K \searrow K$. $\quad\square$

Lemma 4.1.2. *Suppose that a finite simplicial complex K collapses to a subcomplex L and let M be another finite simplicial complex. Then $K * M \searrow L * M$.*

Proof. Let τ_1 be a maximal simplex of $M_1 = M$. Since $K \searrow L$, it is easy to see that $K * M \searrow N_1 = (K * M) \smallsetminus \{\sigma\tau_1 \mid \sigma \in K \smallsetminus L\}$. If τ_2 is a maximal simplex of $M_2 = M_1 \smallsetminus \{\tau_1\}$, then $N_1 \searrow N_2 = N_1 \smallsetminus \{\sigma\tau_2 \mid \sigma \in K \smallsetminus L\}$.

In general we take τ_i a maximal simplex in $M_i = M_{i-1} \smallsetminus \{\tau_{i-1}\}$ and then $N_{i-1} \searrow N_i = N_{i-1} \smallsetminus \{\sigma\tau_i \mid \sigma \in K \smallsetminus L\}$. When M_r is just a vertex, $N_r = L * M$ and therefore $K * M \searrow L * M$. $\qquad\square$

Proposition 4.1.3. *If K and L are subcomplexes of a finite simplicial complex, $K \cup L \searrow K$ if and only if $L \searrow K \cap L$.*

Proof. In both cases the simplices removed are those of L which are not in K. It is easy to verify that the elementary collapses can be performed in the same order. $\qquad\square$

Proposition 4.1.4. *If K is a finite simplicial complex, then $K \nearrow\!\!\!\searrow K'$. In fact we can perform all the collapses and expansions involving complexes of dimension at most $n + 1$ where n is the dimension of K. (In this case we say that K $(n + 1)$-deforms to K').*

Proof. We show that something stronger holds, this is true not only for the barycentric subdivision, but for any stellar subdivision αK of K. Suppose σ is a simplex of K and a is a vertex which is not in K. Then $a\dot\sigma \stackrel{e}{\nearrow} a\sigma \searrow \sigma$ (by Lemma 4.1.1). Therefore $a\dot\sigma lk(\sigma) \nearrow a\sigma lk(\sigma) \searrow \sigma lk(\sigma)$ by Lemma 4.1.2 and then, by Proposition 4.1.3,

$$\alpha K = a\dot\sigma lk(\sigma) + \sigma^c \nearrow a\sigma lk(\sigma) + \sigma^c \searrow \sigma lk(\sigma) + \sigma^c = K$$

where αK is the stellar subdivision at the simplex σ. $\qquad\square$

The notion of simple homotopy types was extended to CW-complexes, which constitute a more appropriate setting for this theory. The Whitehead group $Wh(G)$ of a group G is a quotient of the first K-theory group $K_1(\mathbb{Z}(G))$ (see [23]). The Whitehead group $Wh(K)$ of a connected CW-complex K is the Whitehead group of its fundamental group $Wh(\pi_1(K))$, and in the non-connected case, it is the direct sum of the Whitehead groups of its connected components. There is a geometric equivalent definition of $Wh(K)$ in which the underlying set is a quotient of the set of CW-pairs (L, K) such that K is a strong deformation retract of L. If two homotopy equivalent CW-complexes have trivial Whitehead group, then they are simple homotopy equivalent.

For example, if G is a free group, $Wh(G) = 0$. In particular, contractible CW-complexes are simple homotopy equivalent to a point.

4.2 Simple Homotopy Types: The First Main Theorem

The first mathematician who investigated the relationship between finite spaces and simple homotopy types of polyhedra was Osaki [65]. He showed that if $x \in X$ is a beat point, $\mathcal{K}(X)$ collapses to $\mathcal{K}(X \smallsetminus \{x\})$. In particular, if two finite T_0-spaces, X and Y are homotopy equivalent, their associated simplicial complexes, $\mathcal{K}(X)$ and $\mathcal{K}(Y)$, have the same simple homotopy type. However, there exist finite spaces which are not homotopy equivalent but whose associated complexes have the same simple homotopy type. Consider, for instance, the spaces with the following Hasse diagrams.

They are not homotopy equivalent because they are non-homeomorphic minimal finite spaces. However their associated complexes are triangulations of S^1 and therefore, have the same simple homotopy type.

A more interesting example is the following.

Example 4.2.1 (The Wallet). Let W be a finite T_0-space, whose Hasse diagram is the one of Fig. 4.2 below.

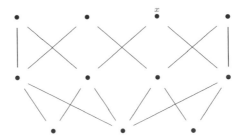

Fig. 4.2 W

This finite space is not contractible since it does not have beat points, but it is not hard to see that $|\mathcal{K}(W)|$ is contractible and therefore, it has the same simple homotopy type as a point. In fact we will deduce from Proposition 4.2.4 that W is a homotopically trivial space, i.e. all its homotopy groups are trivial. This example also shows that the Whitehead Theorem does not hold in the context of finite spaces, not even for homotopically trivial spaces.

We introduce now the notion of a *weak beat point* which generalizes Stong's definition of beat points.

Definition 4.2.2. Let X be a finite T_0-space. We will say that $x \in X$ is a *weak beat point* of X (or a *weak point*, for short) if either $\hat{U}_x$ is contractible or $\hat{F}_x$ is contractible. In the first case we say that x is a *down weak point* and in the second, that x is an *up weak point*.

Note that beat points are in particular weak points since spaces with maximum or minimum are contractible. Since the link $\hat{C}_x = \hat{U}_x \circledast \hat{F}_x$ is a join, we conclude from Proposition 2.7.3 the following

Remark 4.2.3. A point x is a weak point if and only if $\hat{C}_x$ is contractible.

When x is a beat point of X, we have seen that the inclusion $i : X \smallsetminus \{x\} \hookrightarrow X$ is a homotopy equivalence. This is not the case if x is just a weak point. However, a slightly weaker result holds.

Proposition 4.2.4. *Let x be a weak point of a finite T_0-space X. Then the inclusion map $i : X \smallsetminus \{x\} \hookrightarrow X$ is a weak homotopy equivalence.*

Proof. We may suppose that x is a down weak point since the other case follows immediately from this one, considering X^{op} instead of X. Note that $\mathcal{K}(X^{op}) = \mathcal{K}(X)$.

Given $y \in X$, the set $i^{-1}(U_y) = U_y \smallsetminus \{x\}$ has a maximum if $y \neq x$ and is contractible if $y = x$. Therefore $i|_{i^{-1}(U_y)} : i^{-1}(U_y) \to U_y$ is a weak homotopy equivalence for every $y \in X$. Now the result follows from Theorem 1.4.2 applied to the basis like cover given by the minimal basis of X. $\square$

As an application of the last proposition, we verify that the space W defined above, is a non-contractible homotopically trivial space. As we pointed out in Example 4.2.1, W is not contractible since it is a minimal finite space with more than one point. However, it contains a weak point x (see Fig. 4.2), since $\hat{U}_x$ is contractible (see Fig. 4.3). Therefore W is weak

Fig. 4.3 $\hat{U}_x$

homotopy equivalent to $W \smallsetminus \{x\}$ (see Fig. 4.4). Now it is easy to see that this subspace is contractible, because it does have beat points, and one can get rid of them one by one.

Definition 4.2.5. Let X be a finite T_0-space and let $Y \subsetneq X$. We say that X *collapses* to Y by an *elementary collapse* (or that Y *expands* to X by an *elementary expansion*) if Y is obtained from X by removing a weak point. We denote $X \searrow^e Y$ or $Y \nearrow^e X$. In general, given two finite T_0-spaces X and Y, we say that X *collapses* to Y (or Y *expands* to X) if there is a sequence $X = X_1, X_2, \ldots, X_n = Y$ of finite T_0-spaces such that for each $1 \leq i < n$, $X_i \searrow^e X_{i+1}$. In this case we write $X \searrow Y$ or $Y \nearrow X$. Two finite T_0-spaces X and Y are *simple homotopy equivalent* if there is a sequence

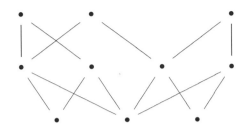

Fig. 4.4 $W \smallsetminus \{x\}$

$X = X_1, X_2, \dots, X_n = Y$ of finite T_0-spaces such that for each $1 \leq i < n$, $X_i \searrow X_{i+1}$ or $X_i \nearrow X_{i+1}$. We denote in this case $X \nearrow\hspace{-0.5em}\searrow Y$, following the same notation that we adopted for simplicial complexes.

Strong collapses studied in Sect. 2.2 are particular cases of collapses.

In contrast with the classical situation, where a simple homotopy equivalence is a special kind of homotopy equivalence, we will see that homotopy equivalent finite spaces are simple homotopy equivalent. In fact this follows almost immediately from the fact that beat points are weak points.

It follows from Proposition 4.2.4 that simple homotopy equivalent finite spaces are weak homotopy equivalent.

In order to prove Theorem 4.2.11, we need some preliminary results. The first one concerns the homotopy type of the associated finite space $\mathcal{X}(K)$ of a simplicial cone K. Suppose $K = aL$ is a cone, i.e. K is the join of a simplicial complex L with a vertex $a \notin L$. Since $|K|$ is contractible, it is clear that $\mathcal{X}(K)$ is homotopically trivial. The following lemma shows that $\mathcal{X}(K)$ is in fact contractible (compare with [70]).

Lemma 4.2.6. *Let $K = aL$ be a finite cone. Then $\mathcal{X}(K)$ is contractible.*

Proof. Define $f : \mathcal{X}(K) \to \mathcal{X}(K)$ by $f(\sigma) = \sigma \cup \{a\}$. This function is order-preserving and therefore continuous.

If we consider the constant map $g : \mathcal{X}(K) \to \mathcal{X}(K)$ that takes all $\mathcal{X}(K)$ into $\{a\}$, we have that $1_{\mathcal{X}(K)} \leq f \geq g$. This proves that the identity is homotopic to a constant map.   □

Recall the construction of the non-Hausdorff mapping cylinder $B(f)$ of a map $f : X \to Y$ between finite T_0-spaces introduced in Sect. 2.8. Denote as before $i : X \to B(f)$ and $j : Y \to B(f)$ the canonical inclusions. It was proved in Lemma 2.8.2 that there is a strong collapse $B(f) \searrow Y$ for any map f. We will show now that under some assumptions on the map f, there is a collapse $B(f) \searrow X$.

Lemma 4.2.7. *Let $f : X \to Y$ be a map between finite T_0-spaces such that $f^{-1}(U_y)$ is contractible for every $y \in Y$. Then $B(f)$ collapses to X.*

Proof. Order the elements $y_1, y_2, \ldots, y_m$ of Y in such a way that $y_r \leq y_s$ implies $r \leq s$ and define $X_r = X \cup \{y_{r+1}, y_{r+2}, \ldots, y_m\} \subseteq B(f)$ for every $0 \leq r \leq m$. Then

$$\hat{U}_{y_r}^{X_{r-1}} = \{x \mid f(x) \leq y_r\}$$

is homeomorphic to $f^{-1}(U_{y_r}^Y)$, which is contractible by hypothesis. Thus $X_{r-1} \searrow^e X_r$ for $1 \leq r \leq m$ and therefore $B(f) = X_0$ collapses to $X = X_m$.
□

Notice that in Definition 4.2.5 it is not explicit that homeomorphic finite T_0-spaces are simple homotopy equivalent. One could have added that to the definition, but it is not needed since it can be deduced from it. If X and Y are disjoint homeomorphic finite T_0-spaces, then we can take a homeomorphism $f : X \to Y$ and the underlying set of $B(f)$ as the union of the disjoint sets X and Y. Then by Lemmas 2.8.2 and 4.2.7, $X \nearrow B(f) \searrow Y$. In the case that X and Y are non-disjoint, one can choose a third space Z homeomorphic to X and Y and disjoint from both of them. Therefore $X \nearrow\searrow Z \nearrow\searrow Y$.

Now we can deduce the following

Lemma 4.2.8. *Homotopy equivalent finite T_0-spaces are simple homotopy equivalent.*

Proof. Suppose $X \overset{he}{\simeq} Y$ and that X_c and Y_c are cores of X and Y. Since beat points are weak points, $X \searrow X_c$ and $Y \searrow Y_c$. On the other hand, X_c and Y_c are homeomorphic and therefore, $X_c \nearrow\searrow Y_c$.
□

As it was observed in Proposition 4.1.4, any finite simplicial complex K has the same simple homotopy type as its barycentric subdivision K'. We prove next an analogous result for finite spaces. Recall that $X' = \mathcal{X}(\mathcal{K}(X))$ denotes the barycentric subdivision of a finite T_0-space X. It is the poset of nonempty chains of X ordered by inclusion. It is shown in [80] and in [35] that there is a weak homotopy equivalence $h : X' \to X$ which takes each chain C to its maximum $\max(C)$. This can be deduced from the proof of the next result.

Proposition 4.2.9. *Let X be a finite T_0-space. Then X and X' are simple homotopy equivalent.*

Proof. Since $B(h) \searrow X$ by Lemma 2.8.2, it suffices to show that the map $h : X' \to X$ satisfies the hypothesis of Lemma 4.2.7. This is clear since $h^{-1}(U_x) = \{C \mid \max(C) \leq x\} = (U_x)' = \mathcal{X}(x\mathcal{K}(\hat{U}_x))$ is contractible by Lemma 4.2.6 (in fact, if a finite T_0-space Y is contractible, so is Y' (see Corollary 5.2.4)).
□

The proof of Proposition 4.2.9 shows that h is a weak homotopy equivalence. Moreover, any map in the hypothesis of Lemma 4.2.7 is a weak homotopy equivalence by Theorem 1.4.2.

Lemma 4.2.10. *Let v be a vertex of a finite simplicial complex K. Then, $lk(v)$ is collapsible if and only if $K \searrow K \searrow v$.*

Proof. By Lemma 4.1.1, $lk(v)$ is collapsible if and only if $st(v) = vlk(v) \searrow lk(v) = st(v) \cap (K \searrow v)$, and this is equivalent to saying that $K = st(v) \cup (K \searrow v) \searrow K \searrow v$ by Proposition 4.1.3. $\qquad\square$

Theorem 4.2.11.

(a) *Let X and Y be finite T_0-spaces. Then, X and Y are simple homotopy equivalent if and only if $\mathcal{K}(X)$ and $\mathcal{K}(Y)$ have the same simple homotopy type. Moreover, if $X \searrow Y$ then $\mathcal{K}(X) \searrow \mathcal{K}(Y)$.*

(b) *Let K and L be finite simplicial complexes. Then, K and L are simple homotopy equivalent if and only if $\mathcal{X}(K)$ and $\mathcal{X}(L)$ have the same simple homotopy type. Moreover, if $K \searrow L$ then $\mathcal{X}(K) \searrow \mathcal{X}(L)$.*

Proof. Let X be a finite T_0-space and assume first that $x \in X$ is a beat point. Then there exists $x' \in X$ and subspaces $Y, Z \subseteq X$ such that $\hat{C}_x = Y \circledast \{x'\} \circledast Z$. The link $lk(x)$ of the vertex x in $\mathcal{K}(X)$ is collapsible, since $lk(x) = \mathcal{K}(\hat{C}_x) = x'\mathcal{K}(Y \circledast Z)$ is a simplicial cone. By Lemma 4.2.10, $\mathcal{K}(X) \searrow \mathcal{K}(X \searrow \{x\})$. In particular, if X is contractible, $\mathcal{K}(X)$ is collapsible.

Now suppose $x \in X$ is a weak point. Then $\hat{C}_x$ is contractible and therefore $lk(x) = \mathcal{K}(\hat{C}_x)$ is collapsible. Again, by Lemma 4.2.10, $\mathcal{K}(X) \searrow \mathcal{K}(X \searrow \{x\})$. We have then proved that $X \searrow Y$ implies $\mathcal{K}(X) \searrow \mathcal{K}(Y)$. In particular, $X \nearrow\!\!\!\!\diagdown Y$ implies $\mathcal{K}(X) \nearrow\!\!\!\!\diagdown \mathcal{K}(Y)$.

Suppose now that K and L are finite simplicial complexes such that $K \searrow^e L$. Then there exist $\sigma \in K$ and a vertex a of K not in σ such that $a\sigma \in K$, $K = L \cup \{\sigma, a\sigma\}$ and $a\sigma \cap L = a\dot\sigma$. It follows that σ is an up beat point of $\mathcal{X}(K)$, and since $\hat{U}_{a\sigma}^{\mathcal{X}(K) \searrow \{\sigma\}} = \mathcal{X}(a\dot\sigma)$, by Lemma 4.2.6, $a\sigma$ is a down weak point of $\mathcal{X}(K) \searrow \{\sigma\}$. Therefore $\mathcal{X}(K) \searrow^e \mathcal{X}(K) \searrow \{\sigma\} \searrow^e \mathcal{X}(K) \searrow \{\sigma, a\sigma\} = \mathcal{X}(L)$. This proves the first part of (b) and the "moreover" part.

Let X, Y be finite T_0-spaces such that $\mathcal{K}(X) \nearrow\!\!\!\!\diagdown \mathcal{K}(Y)$. Then $X' = \mathcal{X}(\mathcal{K}(X)) \nearrow\!\!\!\!\diagdown \mathcal{X}(\mathcal{K}(Y)) = Y'$ and by Proposition 4.2.9, $X \nearrow\!\!\!\!\diagdown Y$. Finally, if K, L are finite simplicial complexes such that $\mathcal{X}(K) \nearrow\!\!\!\!\diagdown \mathcal{X}(L)$, $K' = \mathcal{K}(\mathcal{X}(K)) \nearrow\!\!\!\!\diagdown \mathcal{K}(\mathcal{X}(L)) = L'$ and therefore $K \nearrow\!\!\!\!\diagdown L$ by Proposition 4.1.4. This completes the proof. $\qquad\square$

Corollary 4.2.12. *The functors $\mathcal{K}$, $\mathcal{X}$ induce a one-to-one correspondence between simple homotopy types of finite spaces and simple homotopy types of finite simplicial complexes*

$$\{Finite\ T_0 - Spaces\} \Big/ {\nearrow\!\!\!\!\diagdown} \quad \underset{\mathcal{X}}{\overset{\mathcal{K}}{\rightleftarrows}} \quad \{Finite\ Simplicial\ Complexes\} \Big/ {\nearrow\!\!\!\!\diagdown}$$

The following diagrams illustrate the whole situation.

$$X \overset{he}{\simeq} Y \Longrightarrow X \diagup\!\!\!\!\diagdown Y \Longrightarrow X \overset{we}{\approx} Y$$

$$\big\Updownarrow \qquad\qquad\qquad \big\Updownarrow$$

$$\mathcal{K}(X) \diagup\!\!\!\!\diagdown \mathcal{K}(Y) \Longrightarrow |\mathcal{K}(X)| \overset{we}{\approx} |\mathcal{K}(Y)| \Longleftrightarrow |\mathcal{K}(X)| \overset{he}{\simeq} |\mathcal{K}(Y)|$$

$$\mathcal{X}(K) \overset{he}{\simeq} \mathcal{X}(L) \Longrightarrow \mathcal{X}(K) \diagup\!\!\!\!\diagdown \mathcal{X}(L) \Longrightarrow \mathcal{X}(K) \overset{we}{\approx} \mathcal{X}(L)$$

$$\big\Updownarrow \qquad\qquad\qquad \big\Updownarrow$$

$$K \diagup\!\!\!\!\diagdown L \Longrightarrow |K| \overset{we}{\approx} |L| \Longleftrightarrow |K| \overset{he}{\simeq} |L|$$

The Wallet W satisfies $W \searrow *$, however $W \overset{he}{\not\simeq} *$. Therefore $X \diagup\!\!\!\!\diagdown Y \not\Rightarrow X \overset{he}{\simeq} Y$. Since $|K| \overset{he}{\simeq} |L| \not\Rightarrow K \diagup\!\!\!\!\diagdown L$, $X \overset{we}{\approx} Y \not\Rightarrow X \diagup\!\!\!\!\diagdown Y$. The *Whitehead group* $Wh(X)$ of a finite T_0-space X is $Wh(\pi_1(X))$ if X is connected, and the direct sum of the Whitehead groups of its connected components in general. Therefore $Wh(X) = Wh(|\mathcal{K}(X)|)$. Note that, if $X \overset{we}{\approx} Y$ and their Whitehead group $Wh(X)$ is trivial, then $|\mathcal{K}(X)|$ and $|\mathcal{K}(Y)|$ are simple homotopy equivalent CW-complexes. It follows from Theorem 4.2.11 that $X \diagup\!\!\!\!\diagdown Y$. Thus, we have proved.

Corollary 4.2.13. *Let X, Y be weak homotopy equivalent finite T_0-spaces with trivial Whitehead group. Then $X \diagup\!\!\!\!\diagdown Y$.*

Beat points defined by Stong provide an effective way of deciding whether two finite spaces are homotopy equivalent. The problem becomes much harder when one deals with weak homotopy types instead. There is no easy way to decide whether two finite spaces are weak homotopy equivalent or not. However if two finite T_0-spaces have trivial Whitehead group, then they are weak homotopy equivalent if and only we can obtain one from the other just by adding and removing weak points.

Another immediate consequence of Theorem 4.2.11 is the following

Corollary 4.2.14. *Let X, Y be finite T_0-spaces. If $X \searrow Y$, then $X' \searrow Y'$.*

Note that from Theorem 4.2.11 one also deduces the following well-known fact: If K and L are finite simplicial complexes such that $K \searrow L$, then $K' \searrow L'$.

4.3   Joins, Products, Wedges and Collapsibility

The notion of collapsibility for finite spaces is closely related to the analogous notion for simplicial complexes: We say that a finite T_0-space is *collapsible* if it collapses to a point. Observe that every contractible finite T_0-space is collapsible, however the converse is not true. The Wallet W introduced in Example 4.2.1 is collapsible and non-contractible. Note that if a finite T_0-space X is collapsible, its associated simplicial complex $\mathcal{K}(X)$ is also collapsible. Moreover, if K is a collapsible complex, then $\mathcal{X}(K)$ is a collapsible finite space. Therefore, if X is a collapsible finite space, its subdivision X' is also collapsible.

Remark 4.3.1. Note that if the link $\hat{C}_x$ of a point $x \in X$ is collapsible, $\mathcal{K}(\hat{C}_x)$ is also collapsible and one has that $\mathcal{K}(X) \searrow \mathcal{K}(X \smallsetminus \{x\})$ by Lemma 4.2.10.

Example 4.3.2. Let W be the Wallet, and $\mathbb{C}(W)$ its non-Hausdorff cone. By Remark 4.3.1, $\mathcal{K}(\mathbb{C}(W)) \searrow \mathcal{K}(W)$ but $\mathbb{C}(W)$ does not collapse to W.

Let us consider now a compact contractible polyhedron X with the property that any triangulation of X is non-collapsible. One such a space is the *Dunce Hat* [87]. The Dunce Hat is the space obtained from a triangle by identifying the edges as it is shown in Fig. 4.5.

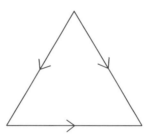

Fig. 4.5 The Dunce hat

This space has a CW-structure with only one 0-cell, one 1-cell and one 2-cell. One way to see that it is contractible is by observing that the attaching map of the 2-cell is a homotopy equivalence $S^1 \to S^1$ and then an easy application of the gluing theorem A.2.5 gives a homotopy equivalence from the 2-dimensional disk to the Dunce Hat. Any triangulation of the Dunce Hat is non-collapsible since each 1-simplex is contained in two or three 2-simplices. Let K be any triangulation of X. The associated finite space $\mathcal{X}(K)$ is homotopically trivial because X is contractible. However, $\mathcal{X}(K)$ is not collapsible since K' is not collapsible. The number of points of the finite space $\mathcal{X}(K)$ constructed in this way is the same as the number of simplices of K. In Chap. 7 we will develop methods for constructing smaller finite models of CW-complexes. In Fig. 7.3 we show a finite space of only 15 points which is homotopically trivial and non-collapsible.

We have therefore the following strict implications in the context of finite spaces:

$$\text{contractible} \Rightarrow \text{collapsible} \Rightarrow \text{homotopically trivial.}$$

Example 4.3.3. The following space X is another example of a collapsible space which is not contractible. It was first considered in [71, Fig. 2].

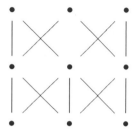

The space $X \cup \{a\}$ below is contractible and collapses to X. Therefore contractibility is not invariant under collapses.

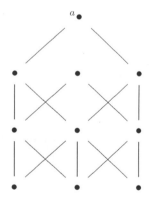

It is known that if K and L are finite simplicial complexes and one of them is collapsible, then $K * L$ is also collapsible (see Lemma 4.1.2). As far as we know the converse of this result is an open problem (see [83, (4.1)]). In the setting of finite spaces, the analogous result and its converse hold.

Proposition 4.3.4. *Let X and Y be finite T_0-spaces. Then $X \circledast Y$ is collapsible if and only if X or Y is collapsible.*

Proof. We proceed as in Proposition 2.7.3, replacing beat points by weak points and deformation retractions by collapses. Note that if x_i is a weak point of X_i, then x_i is also a weak point of $X_i \circledast Y$, since $\hat{C}_{x_i}^{X_i \circledast Y} = \hat{C}_{x_i}^{X_i} \circledast Y$ is contractible by Proposition 2.7.3.

On the other hand, if z_i is a weak point of $X_i \circledast Y_i$ and $z_i \in X_i$, then by Proposition 2.7.3, z_i is a weak point of X_i or Y_i is contractible.   □

By the proof of Proposition 4.3.4 one also has the following

Proposition 4.3.5. *Let X_1, X_2, Y_1, Y_2 be finite T_0-spaces. If $X_1 \diagdown X_2$ and $Y_1 \diagdown Y_2$, $X_1 \circledast Y_1 \diagdown X_2 \circledast Y_2$.*

These are a similar results for products.

Lemma 4.3.6. *Let X and Y be finite T_0-spaces. If $X \searrow A$, $X \times Y \searrow A \times Y$.*

Proof. It suffices to show that if $x \in X$ is a weak point of X, $X \times Y \searrow (X \smallsetminus \{x\}) \times Y$. Suppose without loss of generality that x is a down weak point. If $y \in Y$,

$$\hat{U}_{(x,y)} = \hat{U}_x \times U_y \cup \{x\} \times \hat{U}_y.$$

Let $y_0 \in Y$ be a minimal point. Then $\hat{U}_{(x,y_0)} = \hat{U}_x \times U_{y_0}$ is contractible since each factor is contractible. Therefore, (x, y_0) is a down weak point of $X \times Y$. Now, let y_1 be minimal in $Y \smallsetminus \{y_0\}$. Then $\hat{U}_{(x,y_1)}^{X \times Y \smallsetminus \{(x,y_0)\}} = \hat{U}_x \times U_{y_1}^Y \cup \{x\} \times \hat{U}_{y_1}^Y \smallsetminus \{(x,y_0)\}\} = \hat{U}_x \times U_{y_1}^Y \cup \{x\} \times \hat{U}_{y_1}^{Y \smallsetminus \{y_0\}} = \hat{U}_x \times U_{y_1}^Y$ which again is contractible. Therefore (x, y_1) is a weak point in $X \times Y \smallsetminus \{(x, y_0)\}$. Following this reasoning we remove from $X \times Y$ all the points of the form (x, y) with $y \in Y$.   □

In particular we deduce the following two results.

Proposition 4.3.7. *Let X_1, X_2, Y_1, Y_2 be finite T_0-spaces. If $X_1 \diagdown X_2$ and $Y_1 \diagdown Y_2$, $X_1 \times Y_1 \diagdown X_2 \times Y_2$.*

Proposition 4.3.8. *Let X and Y be collapsible finite T_0-spaces. Then $X \times Y$ is collapsible.*

There is an analogous result to Proposition 4.3.8 for the associated complexes, which relates the collapsibility of $\mathcal{K}(X \times Y)$ with the collapsibility of $\mathcal{K}(X)$ and $\mathcal{K}(Y)$ (see [83]).

The following lemma, was used in the original proof of Theorem 4.2.11 in [8]. The shorter proof we exhibit here does not use this result, but we will need it for the proof of Proposition 4.3.10.

Lemma 4.3.9. *Let L be a subcomplex of a finite simplicial complex K. Let T be a set of simplices of K which are not in L, and let a be a vertex of K which is contained in no simplex of T, but such that $a\sigma$ is a simplex of K for every $\sigma \in T$. Finally, suppose that $K = L \cup \bigcup_{\sigma \in T} \{\sigma, a\sigma\}$ (i.e. the simplices of K are those of L together with the simplices σ and $a\sigma$ for every σ in T). Then $L \nearrow K$.*

Proof. Number the elements $\sigma_1, \sigma_2, \ldots, \sigma_n$ of T in such a way that for every i, j with $i \leq j$, $\#\sigma_i \leq \#\sigma_j$. Here $\#\sigma_k$ denotes the cardinality of σ_k. Define

$K_i = L \cup \bigcup_{j=1}^{i} \{\sigma_j, a\sigma_j\}$ for $0 \leq i \leq n$. Let $\sigma \subsetneq \sigma_i$. If $\sigma \in T$, then $\sigma, a\sigma \in K_{i-1}$, since $\#\sigma < \#\sigma_i$. If $\sigma \notin T$, then $\sigma, a\sigma \in L \subseteq K_{i-1}$. This proves that $a\sigma_i \cap K_{i-1} = a\dot{\sigma}_i$.

By induction, K_i is a simplicial complex for every i, and $K_{i-1} \nearrow K_i$. Therefore $L = K_0 \nearrow K_n = K$.   □

Proposition 4.3.10. *Let (X, x_0) and (Y, y_0) be finite T_0-pointed spaces. Then there exists a weak homotopy equivalence $|\mathcal{K}(X)| \vee |\mathcal{K}(Y)| \to X \vee Y$.*

Proof. Let $\mathcal{K}(X) \vee \mathcal{K}(Y) \subseteq \mathcal{K}(X \vee Y)$ be the simplicial complex which is the union of the complexes $\mathcal{K}(X)$ and $\mathcal{K}(Y)$ identifying the vertices x_0 and y_0. Then $|\mathcal{K}(X)| \vee |\mathcal{K}(Y)|$ is homeomorphic to $|\mathcal{K}(X) \vee \mathcal{K}(Y)|$. The McCord map $\mu_{X \vee Y} : |\mathcal{K}(X \vee Y)| \to X \vee Y$ induces a map $f = \mu_{X \vee Y}i : |\mathcal{K}(X)| \vee |\mathcal{K}(Y)| \to X \vee Y$, where $i : |\mathcal{K}(X)| \vee |\mathcal{K}(Y)| \hookrightarrow |\mathcal{K}(X \vee Y)|$ is the canonical inclusion. In order to prove that f is a weak homotopy equivalence, we only need to prove that i is a homotopy equivalence. We show something stronger: there is a simplicial expansion from $\mathcal{K}(X) \vee \mathcal{K}(Y)$ to $\mathcal{K}(X \vee Y)$.

Take $K = \mathcal{K}(X \vee Y)$ and $L = \mathcal{K}(X) \vee \mathcal{K}(Y)$. Let $a = x_0 = y_0$ and let $T = \{\sigma \in K \mid \sigma \notin L \text{ and } a \notin \sigma\}$. If $\sigma \in T$, then every point of σ is comparable with a, and therefore $a\sigma \in K$. By Lemma 4.3.9, $L \nearrow K$.   □

Corollary 4.3.11. *Let X and Y be finite T_0-spaces. Then $X \vee Y$ is homotopically trivial if and only if both X and Y are.*

Proof. If X and Y are homotopically trivial, the polyhedra $|\mathcal{K}(X)|$ and $|\mathcal{K}(Y)|$ are contractible and therefore $|\mathcal{K}(X)| \vee |\mathcal{K}(Y)|$ is contractible. Thus, $X \vee Y$ is homotopically trivial by Proposition 4.3.10. Conversely, if $X \vee Y$ is homotopically trivial, $|\mathcal{K}(X)| \vee |\mathcal{K}(Y)|$ is contractible and then $|\mathcal{K}(X)|$ and $|\mathcal{K}(Y)|$ are contractible. Therefore, X and Y are homotopically trivial.   □

Suppose that X and Y are finite T_0-spaces and $x_0 \in X$, $y_0 \in Y$ are minimal points. If $X \vee Y$ is collapsible it can be proved by induction that both X and Y are collapsible. If $z \in X \vee Y$ is a weak point, $z \neq \overline{x}_0$ (the class of x_0 in $X \vee Y$) unless $X = *$ or $Y = *$. But the distinguished point $\overline{x}_0 \in X \vee Y$ could be a weak point with $X \neq * \neq Y$ if $x_0 \in X$ or $y_0 \in Y$ is not minimal. It is not known in the general case whether $X \vee Y$ collapsible implies that X and Y are collapsible. However, the converse is false as the next example shows.

Example 4.3.12. The simplicial complex K of Example 11.2.9 is collapsible, and therefore, $\mathcal{X}(K)$ is collapsible. The space $\mathcal{X}(K)$ has a unique weak point σ corresponding to the unique free face of K. Then the union $X = \mathcal{X}(K) \vee \mathcal{X}(K)$ of two copies of $\mathcal{X}(K)$ at $x_0 = \sigma$ is homotopically trivial, but it has no weak points and then it is not collapsible. If $x \in \mathcal{X}(K)$ is distinct from $\overline{x}_0$, $\hat{C}_x^X$ deformation retracts into $\hat{C}_x^{\mathcal{X}(K)}$ which is not contractible. The point $\overline{x}_0 \in X$ is not a weak point either, since its link $\hat{C}_{\overline{x}_0}^X$ is a join of non-connected spaces.

4.4   Simple Homotopy Equivalences: The Second Main Theorem

In this section we prove the second main result of the chapter, which relates simple homotopy equivalences of complexes with simple homotopy equivalences between finite spaces. As in the classical setting, the class of simple homotopy equivalences is generated by the elementary expansions. However, in the context of finite spaces this class is also generated by the *distinguished* maps, which play a key role in this theory.

Recall that a homotopy equivalence $f : |K| \to |L|$ between compact polyhedra is a *simple homotopy equivalence* if it is homotopic to a composition of a finite sequence of maps $|K| \to |K_1| \to \ldots \to |K_n| \to |L|$, each of them an expansion or a homotopy inverse of one [23, 74].

We prove first that homotopy equivalences between finite spaces induce simple homotopy equivalences between the associated polyhedra.

Theorem 4.4.1. *If $f : X \to Y$ is a homotopy equivalence between finite T_0-spaces, then $|\mathcal{K}(f)| : |\mathcal{K}(X)| \to |\mathcal{K}(Y)|$ is a simple homotopy equivalence.*

Proof. Let X_c and Y_c be cores of X and Y. Let $i_X : X_c \to X$ and $i_Y : Y_c \to Y$ be the inclusions and $r_X : X \to X_c$, $r_Y : Y \to Y_c$ retractions of i_X and i_Y such that $i_X r_X \simeq 1_X$ and $i_Y r_Y \simeq 1_Y$.

Since $r_Y f i_X : X_c \to Y_c$ is a homotopy equivalence between minimal finite spaces, it is a homeomorphism. Therefore $\mathcal{K}(r_Y f i_X) : \mathcal{K}(X_c) \to \mathcal{K}(Y_c)$ is an isomorphism and then $|\mathcal{K}(r_Y f i_X)|$ is a simple homotopy equivalence. Since $\mathcal{K}(X) \searrow \mathcal{K}(X_c)$, $|\mathcal{K}(i_X)|$ is a simple homotopy equivalence, and then the homotopy inverse $|\mathcal{K}(r_X)|$ is also a simple homotopy equivalence. Analogously $|\mathcal{K}(i_Y)|$ is a simple homotopy equivalence.

Finally, since $f \simeq i_Y r_Y f i_X r_X$, it follows that

$$|\mathcal{K}(f)| \simeq |\mathcal{K}(i_Y)||\mathcal{K}(r_Y f i_X)||\mathcal{K}(r_X)|$$

is a simple homotopy equivalence.   □

In order to describe the class of simple homotopy equivalences between finite spaces, we will use a kind of map that was already studied in Lemma 4.2.7.

Definition 4.4.2. A map $f : X \to Y$ between finite T_0-spaces is *distinguished* if $f^{-1}(U_y)$ is contractible for each $y \in Y$. We denote by $\mathcal{D}$ the class of distinguished maps.

Note that by the Theorem of McCord 1.4.2, every distinguished map is a weak homotopy equivalence and therefore induces a homotopy equivalence between the associated complexes. We will prove in Theorem 4.4.4 that in fact the induced map is a simple homotopy equivalence. From the proof of

Proposition 4.2.4, it is clear that if $x \in X$ is a down weak point, the inclusion $X \smallsetminus \{x\} \hookrightarrow X$ is distinguished.

Remark 4.4.3. The map $h : X' \to X$ defined by $h(C) = \max(C)$, is distinguished by the proof of Proposition 4.2.9.

Clearly, homeomorphisms are distinguished. However it is not difficult to show that homotopy equivalences are not distinguished in general.

Theorem 4.4.4. *Every distinguished map induces a simple homotopy equivalence.*

Proof. Suppose $f : X \to Y$ is distinguished. Consider the non-Hausdorff mapping cylinder $B(f)$ and the canonical inclusions $i : X \hookrightarrow B(f)$, $j : Y \hookrightarrow B(f)$. Recall that there is a retraction $r : B(f) \to Y$ defined by $r(x) = f(x)$ for every $x \in X$ and that r is a homotopy equivalence (see Sect. 2.8). Then $|\mathcal{K}(f)| = |\mathcal{K}(r)||\mathcal{K}(i)|$. By Lemma 4.2.7 and Theorem 4.2.11, $|\mathcal{K}(i)|$ is an expansion and by Theorem 4.4.1, $|\mathcal{K}(r)|$ is a simple homotopy equivalence. Therefore $|\mathcal{K}(f)|$ is also a simple homotopy equivalence. $\square$

In Proposition 6.2.10 we will prove that Theorem 4.4.4 also holds for a weaker notion of distinguished map: if $f : X \to Y$ is a map between finite T_0 spaces such that $f^{-1}(U_y)$ is homotopically trivial for every $y \in Y$, then f induces a simple homotopy equivalence.

We have already shown that expansions, homotopy equivalences and distinguished maps induce simple homotopy equivalences at the level of complexes. Note that if f, g, h are three maps between finite T_0-spaces such that $fg \simeq h$ and two of them induce simple homotopy equivalences, then so does the third.

Definition 4.4.5. Let $\mathcal{C}$ be a class of continuous maps between topological spaces. We say that $\mathcal{C}$ is *closed* if it satisfies the following homotopy 2-out-of-3 property: For any f, g, h with $fg \simeq h$, if two of the three maps are in $\mathcal{C}$, then so is the third.

Definition 4.4.6. Let $\mathcal{C}$ be a class of continuous maps. The class $\overline{\mathcal{C}}$ *generated* by $\mathcal{C}$ is the smallest closed class containing $\mathcal{C}$.

It is clear that $\overline{\mathcal{C}}$ is always closed under composition and homotopy. The class of simple homotopy equivalences between CW-complexes is closed and it is generated by the elementary expansions. Note that every map in the class $\mathcal{E}$ of elementary expansions between finite spaces induces a simple homotopy equivalence at the level of complexes and therefore the same holds for the maps of $\overline{\mathcal{E}}$. Contrary to the case of CW-complexes, a map between finite spaces which induces a simple homotopy equivalence need not have a homotopy inverse. This is the reason why the definition of $\overline{\mathcal{E}}$ is not as simple as in the setting of complexes. We will prove that $\overline{\mathcal{E}} = \overline{\mathcal{D}}$, the class generated by the distinguished maps.

A map $f : X \to Y$ such that $f^{-1}(F_y)$ is contractible for every y need not be distinguished. However we will show that $f \in \overline{\mathcal{D}}$. We denote by $f^{op} :$ $X^{op} \to Y^{op}$ the map that coincides with f in the underlying sets, and let $\mathcal{D}^{op} = \{f \mid f^{op} \in \mathcal{D}\}$.

Lemma 4.4.7. $\overline{\mathcal{D}^{op}} = \overline{\mathcal{D}}$.

Proof. Suppose that $f : X \to Y$ lies in $\mathcal{D}^{op}$. Consider the following commutative diagram

$$
\begin{array}{ccccc}
X & \xleftarrow{\ h_X\ } & X' = (X^{op})' & \xrightarrow{\ h_{X^{op}}\ } & X^{op} \\
\downarrow{\scriptstyle f} & & \downarrow{\scriptstyle f'} & & \downarrow{\scriptstyle f^{op}} \\
Y & \xleftarrow{\ h_Y\ } & Y' = (Y^{op})' & \xrightarrow{\ h_{Y^{op}}\ } & Y^{op}.
\end{array}
$$

Here, f' denotes the map $\mathcal{X}(\mathcal{K}(f))$. Since $\overline{\mathcal{D}}$ satisfies the 2-out-of-3 property and $h_{X^{op}}, h_{Y^{op}}, f^{op}$ are distinguished by Remark 4.4.3, $f' \in \overline{\mathcal{D}}$. And since h_X, h_Y are distinguished, $f \in \overline{\mathcal{D}}$. This proves that $\overline{\mathcal{D}^{op}} \subseteq \overline{\mathcal{D}}$. The other inclusion follows analogously from the opposite diagram. $\square$

Proposition 4.4.8. $\overline{\mathcal{E}} = \overline{\mathcal{D}}$, *and this class contains all homotopy equivalences between finite* T_0-*spaces.*

Proof. Every expansion of finite spaces is in $\overline{\mathcal{E}}$ because it is a composition of maps in $\mathcal{E}$.

Let $f : X \to Y$ be distinguished. Using the non-Hausdorff mapping cylinder $B(f)$, we deduce that there exist expansions (eventually composed with homeomorphisms) i, j, such that $i \simeq jf$. Therefore $f \in \overline{\mathcal{E}}$.

If $x \in X$ is a down weak point, the inclusion $X \smallsetminus \{x\} \hookrightarrow X$ is distinguished. If x is an up weak point, $X \smallsetminus \{x\} \hookrightarrow X$ lies in $\overline{\mathcal{D}}$ by the previous lemma and therefore $\overline{\mathcal{E}} \subseteq \overline{\mathcal{D}}$.

Suppose now that $f : X \to Y$ is a homotopy equivalence. From the proof of Theorem 4.4.1, $f i_X \simeq i_Y r_Y f i_X$ where i_X, i_Y are expansions and $r_Y f i_X$ is a homeomorphism. This implies that $f \in \overline{\mathcal{E}} = \overline{\mathcal{D}}$. $\square$

We denote by $\mathcal{S} = \overline{\mathcal{E}} = \overline{\mathcal{D}}$ the class of *simple homotopy equivalences* between finite spaces. In the rest of the section we study the relationship between simple homotopy equivalences of finite spaces and simple homotopy equivalences of polyhedra.

Given $n \in \mathbb{N}$ we denote by $K^{(n)}$ the nth barycentric subdivision of K.

Lemma 4.4.9. *Let* $\lambda : K^{(n)} \to K$ *be a simplicial approximation to the identity. Then* $\mathcal{X}(\lambda) \in \mathcal{S}$.

Proof. Since any approximation $K^{(n)} \to K$ to the identity is contiguous to a composition of approximations $K^{(i+1)} \to K^{(i)}$ for $0 \le i < n$ (see

Proposition A.1.6), by Lemma 2.1.3 it suffices to prove the case $n = 1$. Suppose $\lambda : K' \to K$ is a simplicial approximation of $1_{|K|}$. Then $\mathcal{X}(\lambda) :$ $\mathcal{X}(K)' \to \mathcal{X}(K)$ is homotopic to $h_{\mathcal{X}(K)}$, for if $\sigma_1 \subsetneq \sigma_2 \subsetneq \ldots \subsetneq \sigma_m$ is a chain of simplices of K, then $\mathcal{X}(\lambda)(\{\sigma_1, \sigma_2, \ldots, \sigma_m\}) = \{\lambda(\sigma_1), \lambda(\sigma_2), \ldots, \lambda(\sigma_m)\} \subseteq$ $\sigma_m = h_{\mathcal{X}(K)}(\{\sigma_1, \sigma_2, \ldots, \sigma_m\})$. By Remark 4.4.3, it follows that $\mathcal{X}(\lambda) \in \mathcal{S}$. $\qquad\square$

Lemma 4.4.10. *Let* $\varphi, \psi : K \to L$ *be simplicial maps such that* $|\varphi| \simeq |\psi|$. *If* $\mathcal{X}(\varphi) \in \mathcal{S}$, *then* $\mathcal{X}(\psi)$ *also lies in* $\mathcal{S}$.

Proof. There exists an approximation to the identity $\lambda : K^{(n)} \to K$ for some $n \geq 1$, such that $\varphi\lambda$ and $\psi\lambda$ lie in the same contiguity class (see Proposition A.1.6 and Theorem A.1.7). By Proposition 2.1.3, $\mathcal{X}(\varphi)\mathcal{X}(\lambda) = \mathcal{X}(\varphi\lambda) \simeq$ $\mathcal{X}(\psi\lambda) = \mathcal{X}(\psi)\mathcal{X}(\lambda)$. By Lemma 4.4.9, $\mathcal{X}(\lambda) \in \mathcal{S}$ and since $\mathcal{X}(\varphi) \in \mathcal{S}$, it follows that $\mathcal{X}(\psi) \in \mathcal{S}$. $\qquad\square$

Theorem 4.4.11. *Let* $K_0, K_1, \ldots, K_n$ *be finite simplicial complexes and let*

$$|K_0| \xrightarrow{f_0} |K_1| \xrightarrow{f_1} \cdots \xrightarrow{f_{n-1}} |K_n|$$

be a sequence of continuous maps such that for each $0 \leq i < n$ *either*

(1) $f_i = |\varphi_i|$ *where* $\varphi_i : K_i \to K_{i+1}$ *is a simplicial map such that* $\mathcal{X}(\varphi_i) \in \mathcal{S}$ *or*

(2) f_i *is a homotopy inverse of a map* $|\varphi_i|$ *with* $\varphi_i : K_{i+1} \to K_i$ *a simplicial map such that* $\mathcal{X}(\varphi_i) \in \mathcal{S}$.

If $\varphi : K_0 \to K_n$ *is a simplicial map such that* $|\varphi| \simeq f_{n-1}f_{n-2}\ldots f_0$, *then* $\mathcal{X}(\varphi) \in \mathcal{S}$.

Proof. We may assume that f_0 satisfies condition (1). Otherwise we define $\widetilde{K_0} = K_0$, $\widetilde{f_0} = |1_{K_0}| : |\widetilde{K_0}| \to |K_0|$ and then $|\varphi| \simeq f_{n-1}f_{n-2}\ldots f_0\widetilde{f_0}$.

We proceed by induction on n. If $n = 1$, $|\varphi| \simeq |\varphi_0|$ where $\mathcal{X}(\varphi_0) \in \mathcal{S}$ and the result follows from Lemma 4.4.10. Suppose now that $n \geq 1$ and let $K_0, K_1, \ldots, K_n, K_{n+1}$ be finite simplicial complexes and $f_i : |K_i| \to |K_{i+1}|$ maps satisfying conditions (1) or (2), f_0 satisfying condition (1). Let $\varphi :$ $K_0 \to K_{n+1}$ be a simplicial map such that $|\varphi| \simeq f_n f_{n-1}\ldots f_0$. We consider two cases: f_n satisfies condition (1) or f_n satisfies condition (2).

In the first case we define $g : |K_0| \to |K_n|$ by $g = f_{n-1}f_{n-2}\ldots f_0$. Let $\widetilde{g} : K_0^{(m)} \to K_n$ be a simplicial approximation to g and let $\lambda : K_0^{(m)} \to$ K_0 be a simplicial approximation to the identity. Then $|\widetilde{g}| \simeq g|\lambda| =$ $f_{n-1}f_{n-2}\ldots f_1(f_0|\lambda|)$ where $f_0|\lambda| = |\varphi_0\lambda|$ and $\mathcal{X}(\varphi_0\lambda) = \mathcal{X}(\varphi_0)\mathcal{X}(\lambda) \in \mathcal{S}$ by Lemma 4.4.9. By induction, $\mathcal{X}(\widetilde{g}) \in \mathcal{S}$, and then $\mathcal{X}(\varphi_n\widetilde{g}) \in \mathcal{S}$. Since $|\varphi\lambda| \simeq f_n g|\lambda| \simeq f_n|\widetilde{g}| = |\varphi_n\widetilde{g}|$, by Lemma 4.4.10, $\mathcal{X}(\varphi\lambda)$ lies in $\mathcal{S}$. Therefore $\mathcal{X}(\varphi) \in \mathcal{S}$.

In the other case, $|\varphi_n\varphi| \simeq f_{n-1}f_{n-2}\ldots f_0$ and by induction, $\mathcal{X}(\varphi_n\varphi) \in \mathcal{S}$. Therefore $\mathcal{X}(\varphi)$ also lies in $\mathcal{S}$. $\qquad\square$

Theorem 4.4.12.

(a) Let $f : X \to Y$ be a map between finite T_0-spaces. Then f is a simple
homotopy equivalence if and only if $|\mathcal{K}(f)| : |\mathcal{K}(X)| \to |\mathcal{K}(Y)|$ is a simple
homotopy equivalence.
(b) Let $\varphi : K \to L$ be a simplicial map between finite simplicial complexes.
Then $|\varphi|$ is a simple homotopy equivalence if and only if $\mathcal{X}(\varphi)$ is a simple
homotopy equivalence.

Proof. By definition, if $f \in \mathcal{S}$, $|\mathcal{K}(f)|$ is a simple homotopy equivalence.

Let $\varphi : K \to L$ be a simplicial map such that $|\varphi|$ is a simple homotopy
equivalence. Then there exist finite complexes $K = K_0, K_1, \ldots, K_n = L$
and maps $f_i : |K_i| \to |K_{i+1}|$, which are simplicial expansions or homotopy
inverses of simplicial expansions, and such that $|\varphi| \simeq f_{n-1} f_{n-2} \cdots f_0$. By
Theorem 4.2.11, simplicial expansions between complexes induce expansions
between the associated finite spaces and therefore, by Theorem 4.4.11,
$\mathcal{X}(\varphi) \in \mathcal{S}$.

Suppose now that $f : X \to Y$ is a map such that $|\mathcal{K}(f)|$ is a simple
homotopy equivalence. Then $f' = \mathcal{X}(\mathcal{K}(f)) : X' \to Y'$ lies in $\mathcal{S}$. Since $f h_X = h_Y f'$, $f \in \mathcal{S}$.

Finally, if $\varphi : K \to L$ is a simplicial map such that $\mathcal{X}(\varphi) \in \mathcal{S}$, $|\varphi'| : |K'| \to |L'|$ is a simple homotopy equivalence. Here $\varphi' = \mathcal{K}(\mathcal{X}(\varphi))$ is the
barycentric subdivision of φ. Let $\lambda_K : K' \to K$ and $\lambda_L : L' \to L$ be simplicial
approximations to the identities. Then $\lambda_L \varphi'$ and $\varphi \lambda_K$ are contiguous. In
particular $|\lambda_L||\varphi'| \simeq |\varphi||\lambda_K|$ and then $|\varphi|$ is a simple homotopy equivalence.
$\square$

In the setting of finite spaces one has the following strict inclusions

$$\{homotopy\ equivalences\} \subsetneqq \mathcal{S} \subsetneqq \{weak\ equivalences\}.$$

Clearly, if $f : X \to Y$ is a weak homotopy equivalence between finite
T_0-spaces with trivial Whitehead group, $f \in \mathcal{S}$.

4.5   A Simple Homotopy Version of Quillen's Theorem A

Results which carry local information to global information appear frequently
in Algebraic Topology. The Theorem of McCord 1.4.2 roughly states that if
a map is locally a weak homotopy equivalence, then it is a weak homotopy
equivalence (globally). In the following we prove a result of this kind for
simplicial maps and simple homotopy equivalences.

Let K and L be finite simplicial complexes and let $\varphi : K \to L$ be
a simplicial map. Given a simplex $\sigma \in L$, we denote by $\varphi^{-1}(\sigma)$ the full
subcomplex of K spanned by the vertices $v \in K$ such that $\varphi(v) \in \sigma$.

Recall that the simplicial version of Quillen's Theorem A states that if $\varphi : K \to L$ is a simplicial map and $|\varphi|^{-1}(\overline{\sigma})$ is contractible for every simplex $\sigma \in L$, then $|\varphi|$ is a homotopy equivalence (see [69, p. 93]). This result can be deduced from Quillen's Theorem A or from McCord's Theorem (see the proof of Theorem 4.5.2 below). Note that $|\varphi^{-1}(\sigma)| = |\varphi|^{-1}(\overline{\sigma})$. In particular, if $\varphi^{-1}(\sigma)$ is collapsible for every $\sigma \in L$, $|\varphi|$ is a homotopy equivalence. We prove that under this hypothesis, $|\varphi|$ is a simple homotopy equivalence. Theorem 4.5.2 is stated as it appears in the author's Thesis [5]. However, a stronger result holds (see the discussion at the end of the section).

First, we need to state a stronger version of Lemma 4.2.7. We keep the notation we used there.

Lemma 4.5.1. *Let $f : X \to Y$ be a map between finite T_0-spaces such that $f^{-1}(U_y)$ is collapsible for every $y \in Y$. Then $\mathcal{K}(B(f)) \searrow \mathcal{K}(X)$.*

Proof. We follow the proof and notation of Lemma 4.2.7. The set $\hat{U}_{y_r}^{X_{r-1}} = \{x \mid f(x) \leq y_r\}$ is homeomorphic to $f^{-1}(U_{y_r}^Y)$, which is collapsible by hypothesis. Therefore, $\hat{C}_{y_r}^{X_{r-1}}$ is collapsible by Proposition 4.3.4 and, from Remark 4.3.1, $\mathcal{K}(X_{r-1}) \searrow \mathcal{K}(X_r)$. Thus, $\mathcal{K}(B(f)) = \mathcal{K}(X_0)$ collapses to $\mathcal{K}(X) = \mathcal{K}(X_m)$.   □

Theorem 4.5.2. *Let $\varphi : K \to L$ be a simplicial map between finite simplicial complexes. If $\varphi^{-1}(\sigma)$ is collapsible for every simplex σ of L, then $|\varphi|$ is a simple homotopy equivalence.*

Proof. Let $\sigma \in L$. We show first that $\mathcal{X}(\varphi)^{-1}(U_\sigma) = \mathcal{X}(\varphi^{-1}(\sigma))$. Let $\tau \in K$. Then, $\tau \in \mathcal{X}(\varphi^{-1}(\sigma))$ if and only if τ is a simplex of $\varphi^{-1}(\sigma)$. But this is equivalent to saying that for every vertex v of τ, $\varphi(v) \in \sigma$ or, in other words, that $\varphi(\tau) \subseteq \sigma$ which means that $\mathcal{X}(\varphi)(\tau) \leq \sigma$. By Theorem 4.2.11, $\mathcal{X}(\varphi)^{-1}(U_\sigma)$ is collapsible.

By Lemma 4.5.1, $|\mathcal{K}(i)| : |K'| \to |\mathcal{K}(B(\mathcal{X}(\varphi)))|$ is a simple homotopy equivalence, and so is $|\mathcal{K}(j)| : |L'| \to |\mathcal{K}(B(\mathcal{X}(\varphi)))|$, where $i : \mathcal{X}(K) \hookrightarrow B(\mathcal{X}(\varphi))$ and $j : \mathcal{X}(L) \hookrightarrow B(\mathcal{X}(\varphi))$ are the inclusions. Since $|\mathcal{K}(i)| \simeq |\mathcal{K}(j)||\varphi'|$, $|\varphi'|$ is a simple homotopy equivalence and then, so is $|\varphi|$.   □

Surprisingly, the stronger hypothesis in the theorem is not needed (see [6]). Quillen's fiber Lemma 1.4.19 claims that an order preserving map $f : X \to Y$ between finite posets such that $|\mathcal{K}(f^{-1}(U_y))|$ is contractible for every $y \in Y$, induces a homotopy equivalence $|\mathcal{K}(f)| : |\mathcal{K}(X)| \to |\mathcal{K}(Y)|$. Elementary proofs of this result are given in [80] and in [6]. The advantage of the second proof is that it can be easily modified to obtain a simple homotopy version which can in turn be used to obtain stronger versions of several results, like Dowker's Theorem, the Nerve Lemma and the simplicial version of Quillen's Theorem A. In particular, this last claims that if $\varphi : K \to L$ is a simplicial map and $|\varphi|^{-1}(\overline{\sigma})$ is contractible for every $\sigma \in L$, then $|\varphi|$ is a simple homotopy equivalence.

4.6 Simple, Strong and Weak Homotopy in Two Steps

It is easy to prove that if K_1 and K_2 are simple homotopy equivalent finite CW-complexes, there exists a third complex L such that $K_1 \nearrow L \searrow K_2$ (just perform all the expansions at the beginning and then do the collapses in the order they appear in the formal deformation, [23, Exercise 4.D]). This result says that the formal deformation between K_1 and K_2 can be made in two steps, with one expansion first and a collapse after. When CW-complexes are replaced by simplicial complexes or finite spaces, the structure becomes much more rigid, and the result is not so trivial. If K_1 and K_2 are simple homotopy equivalent finite simplicial complexes, there exists a third complex that collapses to K_1 and to a complex $\widetilde{K}_2$ obtained from K_2 by performing a sequence of stellar subdivisions [84, Theorem 5]. In this section we will prove that if X and Y are finite T_0-spaces, there exists a finite T_0-space Z which collapses to both of them. One such space is obtained by considering the *multiple non-Hausdorff mapping cylinder* which is a generalization of the non-Hausdorff mapping cylinder defined in Definition 2.8.1. We will use this construction to prove a similar result for homotopy types and strong collapses. In the same direction we will prove a result about weak homotopy types at the end of the section.

Definition 4.6.1. Let $X_0, X_1, \ldots, X_n$ be a sequence of finite T_0-spaces and let $f_0, f_1, \ldots, f_{n-1}$ be a sequence of maps such that $f_i : X_i \to X_{i+1}$ or $f_i : X_{i+1} \to X_i$. If $f_i : X_i \to X_{i+1}$ we say that f_i *goes right*, and in other case we say that it *goes left*. We define the *multiple non-Hausdorff mapping cylinder* $B(f_0, f_1, \ldots, f_{n-1}; X_0, X_1, \ldots, X_n)$ as follows. The underlying set is the disjoint union $\bigsqcup_{i=0}^{n} X_i$. We keep the given ordering in each copy X_i and for x and y in different copies, we set $x < y$ in either of the following cases:

- If $x \in X_{2i}$, $y \in X_{2i+1}$ and $f_{2i}(x) \leq y$ or $x \leq f_{2i}(y)$.
- If $x \in X_{2i}$, $y \in X_{2i-1}$ and $f_{2i-1}(x) \leq y$ or $x \leq f_{2i-1}(y)$.

Note that the multiple non-Hausdorff mapping cylinder coincides with the ordinary non-Hausdorff mapping cylinder (Definition 2.8.1) when $n = 1$ and the unique map goes right.

Lemma 4.6.2. *Let $B = B(f_0, f_1, \ldots, f_{n-1}, X_0, X_1, \ldots, X_n)$. If f_0 goes right or if f_0 goes left and it lies in $\mathcal{D}^{op}$, then $B \searrow B \smallsetminus X_0$.*

Proof. If f_0 goes right, $B(f_0)$ strongly collapses to X_1 by Lemma 2.8.2. Since the points of X_0 are not comparable with the points of $X_2, X_3, \ldots X_n$, the same elementary collapses can be performed in B. Then $B \searrow B \smallsetminus X_0$.

Now, if f_0 goes left and $f_0^{op} \in \mathcal{D}$, then $B(f_0^{op}) \searrow X_1^{op}$ by Lemma 4.2.7. Thus, $B(f_0^{op})^{op} \searrow X_1$. On the other hand, $B(f_0^{op}) = B(f_0^{op}; X_1^{op}, X_0^{op}) = B(f_0; X_0, X_1)^{op}$ and then $B(f_0; X_0, X_1) \searrow X_1$. By the same argument as before, $B \searrow B \smallsetminus X_0$. $\square$

The following remark is an easy consequence of the definition.

Remark 4.6.3.

$$B(f_1, f_2, \ldots, f_{n-1}; X_1, X_2, \ldots, X_n)^{op}$$

$$= B(f_0^{op}, f_1^{op}, \ldots, f_{n-1}^{op}; X_0^{op}, X_1^{op}, \ldots X_n^{op}) \searrow X_0^{op}.$$

Lemma 4.6.4. *Let $B = B(f_0, f_1, \ldots, f_{n-1}, X_0, X_1, \ldots, X_n)$. Suppose that*
$f_{2i} \in \mathcal{D}$ *if f_{2i} goes right.*
$f_{2i} \in \mathcal{D}^{op}$ *if f_{2i} goes left.*
$f_{2i+1} \in \mathcal{D}^{op}$ *if f_{2i+1} goes right.*
$f_{2i+1} \in \mathcal{D}$ *if f_{2i+1} goes left.*
Then $B \searrow X_n$. If in addition n is even, $B \searrow X_0$.

Proof. By Lemma 4.6.2, $B \searrow B \smallsetminus X_0$. By the previous remark,

$$B \smallsetminus X_0 = B(f_1^{op}, f_2^{op}, \ldots, f_{n-1}^{op}; X_1^{op}, X_2^{op}, \ldots, X_n^{op})^{op}.$$

By induction $B(f_1^{op}, f_2^{op}, \ldots, f_{n-1}^{op}; X_1^{op}, X_2^{op}, X_n^{op}) \searrow X_n^{op}$. Therefore $B \searrow B \smallsetminus X_0 \searrow X_n$.

If n is even, $B = B(f_{n-1}, f_{n-2}, \ldots, f_0; X_n, X_{n-1}, \ldots, X_0) \searrow X_0$. $\square$

Theorem 4.6.5. *Let X and Y be simple homotopy equivalent finite T_0-spaces. Then there exists a finite T_0-space Z that collapses to both X and (a copy of) Y.*

Proof. If $X \nearrow\searrow Y$, there exists a sequence of elementary expansions and collapses from X to Y. An elementary expansion $X_i \nearrow X_{i+1}$ induces an inclusion map $X_i \hookrightarrow X_{i+1}$ which lies in $\mathcal{D}$ or $\mathcal{D}^{op}$ depending on if the weak point removed is a down weak point or an up weak point. In particular, there exists a sequence $X = X_0, X_1, X_2, \ldots, X_n = Y$ of finite T_0-spaces and a sequence $f_0, f_1, \ldots, f_{n-1}$ of maps such that $f_i : X_i \to X_{i+1}$ or $f_i : X_{i+1} \to X_i$ and $f_i \in \mathcal{D} \cup \mathcal{D}^{op}$ for every $0 \le i \le n-1$. Adding identities if needed, we can assume that the maps are in the hypothesis of Lemma 4.6.4, and the result follows. $\square$

Proposition 4.6.6. *Let X and Y be homotopy equivalent finite T_0-spaces. Then there exists a finite T_0-space Z that contains both X and (a copy of) Y as strong deformations retracts.*

Proof. The idea is essentially to repeat the proof made for simple homotopy types. We can say that a map $f : X \to Y$ between finite T_0-spaces is *strongly distinguished* if $f^{-1}(U_y)$ has a maximum for every $y \in Y$. Following the proof of Lemma 4.2.7 it is easy to see that if $f : X \to Y$ is strongly distinguished, $B(f) \searrow X$. Now we just replace in Lemmas 4.6.2 and 4.6.4 the class $\mathcal{D}$ by the class of strongly distinguished maps and the collapses by strong collapses.

With these versions of the lemmas, the proof of the proposition is similar to that of Theorem 4.6.5.   □

This result mirrors that for general spaces using the classical mapping cylinder (see [38, Corollary 0.21]). Of course for finite spaces an additional property holds: if X and Y are two homotopy equivalent finite T_0-spaces, there exists a third space Z which is a strong deformation retract of both X and (a copy of) Y. This space Z can be taken to be the core of X.

We have mentioned in previous chapters that if X and Y are two weak homotopy equivalent topological spaces, there exists a third space Z and weak homotopy equivalences $X \leftarrow Z \to Y$. The following is a version of that result for finite spaces.

Proposition 4.6.7. *Let X and Y be two weak homotopy equivalent finite spaces. Then there exists a third finite space Z and weak homotopy equivalences $X \leftarrow Z \to Y$.*

Proof. We may assume that X and Y are T_0. Let $f : |\mathcal{K}(X)| \to |\mathcal{K}(Y)|$ be a homotopy equivalence and let $\varphi : \mathcal{K}(X)^{(n)} \to \mathcal{K}(Y)$ be a simplicial approximation to f. Then $|\varphi|$ is also a homotopy equivalence an therefore $\mathcal{X}(\varphi) : X^{(n+1)} \to Y'$ is a weak homotopy equivalence. Here $X^{(n+1)} = (X^{(n)})'$ denotes the $(n+1)$th barycentric subdivision of X. Since there are weak homotopy equivalences $Y' \to Y$ and $X^{(m+1)} \to X^m$ for every $0 \le m \le n$, there exist weak homotopy equivalences $X \leftarrow X^{(n+1)} \to Y$.   □

The idea of considering iterated barycentric subdivisions of finite spaces together with the simplicial approximation theorem appears in a paper of Hardie and Vermeulen [35] (see also [52, Theorem 8.4]). Similar ideas are later used to construct finite analogues of the complex number multiplication $S^1 \times S^1 \to S^1$ and the Hopf map $S^3 \to S^2$ in [36] (other examples can be found in [34, 37]).

Chapter 5
Strong Homotopy Types

The notion of collapse of finite spaces is directly connected with the concept
of simplicial collapse. In Chap. 2 we studied the notion of elementary strong
collapse which is the fundamental move that describes homotopy types of
finite spaces. In this chapter we will define the notion of *strong collapse* of
simplicial complexes which leads to *strong homotopy types* of complexes. This
notion corresponds to the homotopy types of the associated finite spaces, but
we shall see that it also arises naturally from the concept of contiguity classes.

Strong homotopy types of simplicial complexes have a beautiful characterization which is similar to the description of homotopy types of finite spaces
given by Stong.

Most of the results of this chapter are included in the article [11]. However,
the paper contains many more motivations and applications. The reader interested in a more complete exposition of the theory of strong homotopy
types is encouraged to consult [11].

5.1   A Simplicial Notion of Homotopy

Given a vertex v in a simplicial complex K, we denote by $K \smallsetminus v$ the full
subcomplex of K spanned by the vertices different from v. This is often
called the *deletion* of v. Recall that the link $lk(v)$ is the subcomplex of $K \smallsetminus v$
whose simplices are those $\sigma \in K \smallsetminus v$ such that $\sigma \cup \{v\} \in K$.

If we want to study how the homotopy type of K changes when we remove
a vertex v, it is very useful to analyze the subcomplex $lk(v)$. For instance, if
$|lk(v)|$ is contractible, then $|K|$ and $|K \smallsetminus v|$ have the same homotopy type.
This easily follows from the gluing theorem A.2.5 (see Proposition A.2.6 for a
proof). If $lk(v)$ is collapsible, $st(v) = v(lk(v)) \searrow lk(v) = st(v) \cap (K \smallsetminus v)$ and
therefore $K = st(v) \cup (K \smallsetminus v) \searrow K \smallsetminus v$. The notion of *non-evasive complex*
(see [11,41] for example) is also connected with the relationship among $lk(v)$,

J.A. Barmak, *Algebraic Topology of Finite Topological Spaces and*   73
Applications, Lecture Notes in Mathematics 2032,
DOI 10.1007/978-3-642-22003-6_5, © Springer-Verlag Berlin Heidelberg 2011

K and $K \smallsetminus v$. The following definition, motivated by the notion of beat point, is related to these ideas.

Definition 5.1.1. Let K be a finite simplicial complex and let $v \in K$ be a vertex. We say that there is an *elementary strong collapse* from K to $K \smallsetminus v$ if $lk(v)$ is a simplicial cone $v'L$. In this case we say that v is *dominated* (by v') and we write $K \searrow\!\!\!^e K \smallsetminus v$ (Fig. 5.1). There is a *strong collapse* from a complex K to a subcomplex L if there exists a sequence of elementary strong collapses that starts in K and ends in L. In this case we write $K \searrow\!\!\!\searrow L$. The inverse of a strong collapse is a *strong expansion* and two finite complexes K and L have the same *strong homotopy type* if there is a sequence of strong collapses and strong expansions that starts in K and ends in L.

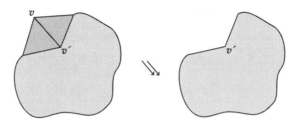

Fig. 5.1 An elementary strong collapse. The vertex v is dominated by v'

Remark 5.1.2. Isomorphic complexes have the same strong homotopy type. Let K be a finite simplicial complex and let $v \in K$ be a vertex. Let v' be a vertex which is not in K and consider the complex $L = K + v'st_K(v) = K \smallsetminus v + v'vlk_K(v)$. Since $lk_L(v') = vlk_K(v)$, $L \searrow\!\!\!\searrow K$. Moreover, by symmetry $L \searrow\!\!\!\searrow L \smallsetminus v = \widetilde{K}$. Clearly, there is an isomorphism $K \to \widetilde{K}$ which sends v to v' and fixes the other vertices. Thus, if K_1 and K_2 are isomorphic simplicial complexes, we can obtain a third complex K_3 whose vertices are different from the vertices of K_1 and K_2 and such that K_i and K_3 have the same strong homotopy type for $i = 1, 2$.

If $v \in K$ is dominated, $lk(v)$ is collapsible and in particular K collapses to $K \smallsetminus v$. Thus, the usual notion of collapse is weaker than the notion of strong collapse.

Remark 5.1.3. A vertex v is dominated by a vertex $v' \neq v$ if and only if every maximal simplex that contains v also contains v'.

We will prove that this notion of collapse corresponds exactly to the notion of strong collapse of finite spaces (i.e. strong deformation retracts).

If two simplicial maps $\varphi, \psi : K \to L$ lie in the same contiguity class, we will write $\varphi \sim \psi$. It is easy to see that if $\varphi_1, \varphi_2 : K \to L$, $\psi_1, \psi_2 : L \to M$ are simplicial maps such that $\varphi_1 \sim \varphi_2$ and $\psi_1 \sim \psi_2$, then $\psi_1\varphi_1 \sim \psi_2\varphi_2$.

Definition 5.1.4. A simplicial map $\varphi : K \to L$ is a *strong equivalence* if there exists $\psi : L \to K$ such that $\psi\varphi \sim 1_K$ and $\varphi\psi \sim 1_L$. If there is a strong equivalence $\varphi : K \to L$ we write $K \sim L$.

The relation $\sim$ is clearly an equivalence relation.

Definition 5.1.5. A finite simplicial complex K is a *minimal complex* if it has no dominated vertices.

Proposition 5.1.6. *Let K be a minimal complex and let $\varphi : K \to K$ be simplicial map which lies in the same contiguity class as the identity. Then φ is the identity.*

Proof. We may assume that φ is contiguous to 1_K. Let $v \in K$ and let $\sigma \in K$ be a maximal simplex such that $v \in \sigma$. Then $\varphi(\sigma) \cup \sigma$ is a simplex, and by the maximality of σ, $\varphi(v) \in \varphi(\sigma) \cup \sigma = \sigma$. Therefore, every maximal simplex which contains v, also contains $\varphi(v)$. Hence, $\varphi(v) = v$, since K is minimal. $\square$

Corollary 5.1.7. *A strong equivalence between minimal complexes is an isomorphism.*

Proposition 5.1.8. *Let K be a finite simplicial complex and $v \in K$ a vertex dominated by v'. Then the inclusion $i : K \smallsetminus v \hookrightarrow K$ is a strong equivalence. In particular, if two complexes K and L have the same strong homotopy type, then $K \sim L$.*

Proof. Define a vertex map $r : K \to K \smallsetminus v$ which is the identity on $K \smallsetminus v$ and such that $r(v) = v'$. If $\sigma \in K$ is a simplex with $v \in \sigma$, consider $\sigma' \supseteq \sigma$ a maximal simplex. Therefore $v' \in \sigma'$ and $r(\sigma) = \sigma \cup \{v'\} \smallsetminus \{v\} \subseteq \sigma'$ is a simplex of $K \smallsetminus v$. Moreover $ir(\sigma) \cup \sigma = \sigma \cup \{v'\} \subseteq \sigma'$ is a simplex of K. This proves that r is simplicial and that ir is contiguous to 1_K. Therefore, i is a strong equivalence. $\square$

Definition 5.1.9. A *core* of a finite simplicial complex K is a minimal subcomplex $K_0 \subseteq K$ such that $K \searrow K_0$.

Theorem 5.1.10. *Every complex has a core and it is unique up to isomorphism. Two finite simplicial complexes have the same strong homotopy type if and only if their cores are isomorphic.*

Proof. A core of a complex can be obtained removing dominated points one by one. If K_1 and K_2 are two cores of K, they have the same strong homotopy type and by Proposition 5.1.8, $K_1 \sim K_2$. Since they are minimal, by Corollary 5.1.7 they are isomorphic.

Let K, L be two finite complexes. If they have the same strong homotopy type, then also their cores K_0 and L_0 do. As above, we conclude that K_0 and L_0 are isomorphic.

Conversely, If K_0 and L_0 are isomorphic, they have the same strong homotopy type by Remark 5.1.2 and then K and L have the same strong homotopy type. $\square$

The uniqueness of cores is a fundamental property that distinguishes strong homotopy types from simple homotopy types. A simplicial complex can collapse to non-isomorphic subcomplexes, each of them without any free face. However if a complex strongly collapses to two minimal complexes, they must be isomorphic. The uniqueness of cores is also proved in [48] where the notion of strong collapse appears with the name of LC-reduction. That paper is exclusively devoted to prove that result from a combinatorial viewpoint. In contrast, Theorem 5.1.10 is motivated by Stong's topological ideas.

If K and L are two complexes such that $K \sim L$ and $K_0 \subseteq K$, $L_0 \subseteq L$ are their cores, then $K_0 \sim L_0$ and therefore K_0 and L_0 are isomorphic. Hence, we deduce the following

Corollary 5.1.11. *Two finite simplicial complexes K and L have the same strong homotopy type if and only if $K \sim L$.*

Example 5.1.12. The following homogeneous 2-complex is collapsible (moreover it is non-evasive [83]). However, it is a minimal complex and therefore it does not have the strong homotopy type of a point.

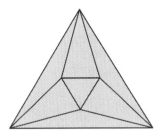

Example 5.1.13. In contrast to the case of simple homotopy types, a complex and its barycentric subdivision need not have the same strong homotopy type. The boundary of a 2-simplex and its barycentric subdivision are minimal non-isomorphic complexes, therefore they do not have the same strong homotopy type.

Proposition 5.1.14. *Strong equivalences are simple homotopy equivalences.*

Proof. Let $\varphi : K \to L$ be a strong equivalence. Let K_0 be a core of K and L_0 a core of L. Then the inclusion $i : K_0 \hookrightarrow K$ is a strong equivalence and there exists a strong equivalence $r : L \to L_0$ which is a homotopy inverse of the inclusion $L_0 \hookrightarrow L$. Since K_0 and L_0 are minimal complexes, the strong equivalence $r\varphi i$ is an isomorphism. Therefore, $|i|, |r|$ and $|r\varphi i|$ are simple homotopy equivalences, and then so is $|\varphi|$. □

Definition 5.1.15. A complex is said to be *strong collapsible* if it strong collapses to a point or equivalently if it has the strong homotopy type of a point.

Recall that it is not known whether $K * L$ is collapsible only if one of K or L is, but the analogous result is true for strong collapsibility.

Proposition 5.1.16. *Let K and L be two finite simplicial complexes. Then, $K * L$ is strong collapsible if and only if K or L is strong collapsible.*

Proof. Suppose v is a dominated vertex of K. Then $lk_K(v)$ is a cone and therefore $lk_{K*L}(v) = lk_K(v) * L$ is a cone. Therefore v is also dominated in $K * L$. Thus, if K strong collapses to a vertex v_0, $K * L \searrow v_0 L \searrow v_0$.

Conversely, assume $K*L$ is strong collapsible. Let $v \in K*L$ be a dominated point and suppose without loss of generality that $v \in K$. Then $lk_{K*L}(v) = lk_K(v) * L$ is a cone. Therefore $lk_K(v)$ is a cone or L is a cone. If L is a cone, it is strong collapsible and we are done. Suppose then that $lk_K(v)$ is a cone. Since $(K \smallsetminus v) * L = (K * L) \smallsetminus v$ is strong collapsible, by induction $K \smallsetminus v$ or L is strong collapsible and since $K \searrow K \smallsetminus v$, K or L is strong collapsible. $\square$

5.2 Relationship with Finite Spaces and Barycentric Subdivisions

In this section we will study the relationship between strong homotopy types of simplicial complexes and homotopy types of finite spaces. After the first result it will be clear that if a finite space is contractible, then so is its barycentric subdivision. The converse of this result, however, is not trivial. We will prove an analogous statement for strong collapsibility of complexes and then we will use it to prove the finite space version.

The following result is a direct consequence of Propositions 2.1.2 and 2.1.3.

Theorem 5.2.1.

(a) *If two finite T_0-spaces are homotopy equivalent, their associated complexes have the same strong homotopy type.*

(b) *If two finite complexes have the same strong homotopy type, the associated finite spaces are homotopy equivalent.*

Proof. Suppose $f : X \to Y$ is a homotopy equivalence between finite T_0-spaces with homotopy inverse $g : Y \to X$. Then by Proposition 2.1.2, $\mathcal{K}(g)\mathcal{K}(f) \sim 1_{\mathcal{K}(X)}$ and $\mathcal{K}(f)\mathcal{K}(g) \sim 1_{\mathcal{K}(Y)}$. Thus, $\mathcal{K}(X) \sim \mathcal{K}(Y)$.

If K and L are complexes with the same strong homotopy type, there exist $\varphi : K \to L$ and $\psi : L \to K$ such that $\psi\varphi \sim 1_K$ and $\varphi\psi \sim 1_L$. By Proposition 2.1.3, $\mathcal{X}(\varphi) : \mathcal{X}(K) \to \mathcal{X}(L)$ is a homotopy equivalence with homotopy inverse $\mathcal{X}(\psi)$. $\square$

In fact, we will give a more precise result:

Theorem 5.2.2.

(a) *Let X be a finite T_0-space and let $Y \subseteq X$. If $X \searrow Y$, $\mathcal{K}(X) \searrow \mathcal{K}(Y)$.*

(b) *Let K be a finite simplicial complex and let $L \subseteq K$. If $K \searrow L$, $\mathcal{X}(K) \searrow \mathcal{X}(K)$.*

Proof. If $x \in X$ is a beat point, there exists a point $x' \in X$ and subspaces A, B such that $\hat{C}_x = A \circledast \{x'\} \circledast B$. Then $lk(x) = \mathcal{K}(\hat{C}_x) = x'\mathcal{K}(A \circledast B)$ is a simplicial cone. Therefore, $\mathcal{K}(X) \searrow \mathcal{K}(X) \smallsetminus x = \mathcal{K}(X \smallsetminus \{x\})$.

If K is a finite complex and $v \in K$ is such that $lk(v) = aL$ is a simplicial cone, we define $r : \mathcal{X}(K) \to \mathcal{X}(K \smallsetminus v)$ as follows:

$$r(\sigma) = \begin{cases} a\sigma \smallsetminus \{v\} & \text{if } v \in \sigma \\ \sigma & \text{if } v \notin \sigma \end{cases}$$

Clearly r is a well defined order preserving map. Denote $i : \mathcal{X}(K \smallsetminus v) \hookrightarrow \mathcal{X}(K)$ the inclusion and define $f : \mathcal{X}(K) \to \mathcal{X}(K)$,

$$f(\sigma) = \begin{cases} a\sigma & \text{if } v \in \sigma \\ \sigma & \text{if } v \notin \sigma \end{cases}$$

Then $ir \leq f \geq 1_{\mathcal{X}(K)}$ and both ir and f are the identity on $\mathcal{X}(K \smallsetminus v)$. Therefore $ir \simeq 1_{\mathcal{X}(K)}$ rel $\mathcal{X}(K \smallsetminus v)$ and then $\mathcal{X}(K) \searrow \mathcal{X}(K \smallsetminus v)$ by Corollary 2.2.5.   □

Example 5.2.3. The complex $\mathcal{K}(W)$ associated to the Wallet (see Fig. 5.2) is a triangulation of the 2-dimensional disk D^2 which is collapsible because W is collapsible, but which is not strong collapsible since W' is not contractible.

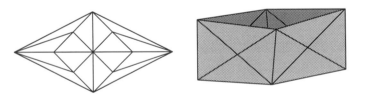

Fig. 5.2 The geometric realization of $\mathcal{K}(W)$

Corollary 5.2.4. *If X is a contractible finite T_0-space, so is X'.*

Proof. If X is contractible, $X \searrow *$, then $\mathcal{K}(X) \searrow *$ and therefore $X' = \mathcal{X}(\mathcal{K}(X)) \searrow *$.   □

In general, for a finite T_0-space X, X and X' do not have the same homotopy type. Nevertheless, we will prove the converse of Corollary 5.2.4, which implies that X is contractible if and only if its barycentric subdivision X' is contractible. In particular, not only W' is non-contractible, but all the iterated barycentric subdivisions of W.

It is not true that if X is a minimal finite space, then so is X'. The barycentric subdivision W' of the Wallet is not a minimal finite space, although W is.

Proposition 5.2.5. *Let X be a finite T_0-space. Then X is a minimal finite space if and only if $\mathcal{K}(X)$ is a minimal simplicial complex.*

Proof. If X is not minimal, it has a beat point x and then $\mathcal{K}(X) \searrow \mathcal{K}(X \smallsetminus \{x\})$ by Theorem 5.2.2. Therefore $\mathcal{K}(X)$ is not minimal.

Conversely, suppose $\mathcal{K}(X)$ is not minimal. Then it has a dominated vertex x. Suppose $lk(x) = x'L$ for some $x' \in X$, $L \subseteq \mathcal{K}(X)$. In particular, if $y \in X$ is comparable with x, $y \in lk(x)$ and then $yx' \in lk(x)$. Thus, any point comparable with x is also comparable with x'. By Proposition 1.3.9, X is not minimal. $\qquad\square$

Theorem 5.2.6. *Let K be a finite simplicial complex. Then K is strong collapsible if and only if K' is strong collapsible.*

Proof. If $K \searrow *$, then $\mathcal{X}(K) \searrow *$ and $K' = \mathcal{K}(\mathcal{X}(K)) \searrow *$ by Theorem 5.2.2. Suppose now that K is a complex and that $K' \searrow *$. Let L be a core of K. Then $K \searrow L$ and by Theorem 5.2.2, $K' \searrow L'$. Therefore L is minimal and L' is strong collapsible. Let $L_0 = L', L_1, L_2, ..., L_n = *$ be a sequence of subcomplexes of L' such that there is an elementary strong collapse from L_i to L_{i+1} for every $0 \le i < n$. We will prove by induction in i that $L_i \subseteq L'$ contains as vertices all the barycenters of the 0-simplices and of the maximal simplices of L.

Let $\sigma = \{v_0, v_1, \ldots, v_k\}$ be a maximal simplex of L. By induction, the barycenter $b(\sigma)$ of σ is a vertex of L_i. We claim that $lk_{L_i} b(\sigma)$ is not a cone. If σ is a 0-simplex, that link is empty, so we assume σ has positive dimension. Since $b(v_j)b(\sigma)$ is a simplex of L, $b(v_j) \in L_i$ by induction and L_i is a full subcomplex of L, then $b(v_j) \in lk_{L_i} b(\sigma)$ for every $0 \le j \le k$. Suppose $lk_{L_i} b(\sigma)$ is a cone. In particular, there exists $\sigma' \in L$ such that $b(\sigma') \in lk_{L_i} b(\sigma)$ and moreover $b(\sigma')b(v_j) \in lk_{L_i} b(\sigma)$ for every j. Since σ is a maximal simplex, $\sigma' \subsetneq \sigma$ and $v_j \in \sigma'$ for every j. Then $\sigma \subseteq \sigma'$, which is a contradiction. Hence, $b(\sigma)$ is not a dominated vertex of L_i and therefore, $b(\sigma) \in L_{i+1}$.

Let $v \in L$ be a vertex. By induction, $b(v) \in L_i$. As above, if v is a maximal simplex of L, $lk_{L_i} b(v) = \emptyset$. Suppose v is not a maximal simplex of L. Let $\sigma_0, \sigma_1, \ldots, \sigma_k$ be the maximal simplices of L which contain v. By induction $b(\sigma_j) \in L_i$ for every $0 \le j \le k$, and since $L_i \subseteq L$ is full, $b(\sigma_j) \in lk_{L_i} b(v)$. Suppose that $lk_{L_i} b(v)$ is cone. Then there exists $\sigma \in K$ such that $b(\sigma) \in lk_{L_i} b(v)$ and moreover, $b(\sigma)b(\sigma_j) \in lk_{L_i} b(v)$ for every j. In particular, $v \subsetneq \sigma$ and $\sigma \subseteq \sigma_j$ for every j. Let $v' \in \sigma$, $v' \ne v$. Then v' is contained in every maximal simplex which contains v. This contradicts the minimality of L. Therefore $b(v)$ is not dominated in L_i, which proves that $b(v) \in L_{i+1}$.

Finally, $L_n = *$ contains all the barycenters of the vertices of L. Thus, $L = *$ and K is strong collapsible. $\qquad\square$

Corollary 5.2.7. *Let X be a finite T_0-space. Then X is contractible if and only if X' is contractible.*

Proof. By Corollary 5.2.4, it only remains to show that if X' is contractible, so is X. Let $Y \subseteq X$ be a core of X. Then by Theorem 5.2.2, $X' \searrow Y'$. If X' is contractible, so is Y'. Again by Theorem 5.2.2, $\mathcal{K}(Y') = \mathcal{K}(Y)'$ is strong collapsible. By Theorem 5.2.6, $\mathcal{K}(Y)$ is strong collapsible. By Proposition 5.2.5, $\mathcal{K}(Y)$ is a minimal complex and therefore $\mathcal{K}(Y) = *$. Then Y is just a point, so X is contractible.   □

Corollary 5.2.8.

1. *A finite T_0-space X is contractible if and only if $\mathcal{K}(X)$ is strong collapsible.*
2. *A finite simplicial complex K is strong collapsible if and only if $\mathcal{X}(K)$ is contractible.*

5.3 Nerves of Covers and the Nerve of a Complex

We introduce an application which transforms a simplicial complex into another complex with the same homotopy type. This construction was previously considered by Grünbaum in [32] (see also [47]) but we arrived to it independently when studying the Čech cohomology of finite spaces. We will prove that this application can be used to obtain the core of a simplicial complex.

Recall that if $\mathcal{U} = \{U_i\}_{i \in I}$ is a cover of a set X, the *nerve* of $\mathcal{U}$ is the simplicial complex $N(\mathcal{U})$ whose simplices are the finite subsets $I' \subseteq I$ such that $\bigcap_{i \in I'} U_i$ is nonempty.

The following result of McCord relates nerves of coverings and weak homotopy equivalences.

Theorem 5.3.1 (McCord). *Let X be a topological space and let $\mathcal{U}$ be an open cover of X such that any point of X is contained in finitely many elements of $\mathcal{U}$. If any intersection of elements of $\mathcal{U}$ is empty or homotopically trivial, there exists a weak homotopy equivalence $|N(\mathcal{U})| \to X$.*

The proof of this result uses Theorem 1.4.2 and can be found in [56, Theorem 2]. This result is closely related to the so called Nerve Lemma [16, Theorem 10.6]. The Nerve Lemma claims that if a finite simplicial complex K is covered by a finite collection of subcomplexes $\mathcal{V}$ such that any intersection of members of $\mathcal{V}$ is empty or contractible, then $|N(\mathcal{V})| \simeq |K|$. The Nerve Lemma follows immediately from Theorem 5.3.1 by considering the space $\mathcal{X}(K)$ and the cover $\mathcal{U} = \{\mathcal{X}(L) \mid L \in \mathcal{V}\}$.

If $\mathcal{V} = \{V_j\}_{j \in J}$ is a refinement of a cover $\mathcal{U} = \{U_i\}_{i \in I}$ of a set X (i.e. every member of $\mathcal{V}$ is contained in some element of $\mathcal{U}$), there is a simplicial map $N(\mathcal{V}) \to N(\mathcal{U})$ which is uniquely determined up to homotopy, and sends any vertex $j \in N(\mathcal{V})$ to a vertex $i \in N(\mathcal{U})$ such that $V_j \subseteq U_i$. In fact any two such maps φ, ψ are contiguous. If $J' \subseteq J$ is a simplex of $N(\mathcal{V})$, then the intersection $\bigcap\limits_{i \in \varphi(J') \cup \psi(J')} U_i$ contains $\bigcap\limits_{j \in J'} V_j$ which is nonempty, and therefore $\varphi(J') \cup \psi(J')$ is a simplex of $N(\mathcal{U})$. We call a map $\varphi : N(\mathcal{V}) \to N(\mathcal{U})$ as above a *refinement map*.

Proposition 5.3.2. *If* $\mathcal{U} = \{U_i\}_{i \in I}$ *and* $\mathcal{V} = \{V_j\}_{j \in J}$ *are two finite covers of a set that refine each other,* $N(\mathcal{U})$ *and* $N(\mathcal{V})$ *have the same strong homotopy type.*

Proof. Let $\varphi : N(\mathcal{U}) \to N(\mathcal{V})$ and $\psi : N(\mathcal{V}) \to N(\mathcal{U})$ be two refinement maps. Then $\psi\varphi$ and $1_{N(\mathcal{U})}$ are two refinements maps and therefore they are contiguous. Analogously $\varphi\psi \sim 1_{N(\mathcal{V})}$. Thus, φ is a strong homotopy equivalence. $\qquad\square$

The Čech cohomology of a topological space X is the direct limit

$$\check{H}^n(X) = \operatorname{colim}\ H^n(N(\mathcal{U}))$$

taken over the family of open covers of X preordered by refinement.

It is well known that if X has the homotopy type of a CW-complex, the Čech cohomology coincides with the singular cohomology of X. But this is not true in general. Given a finite space X, we denote by $\mathcal{U}_0$ the open cover given by the minimal open sets of the maximal points of X. Note that $\mathcal{U}_0$ refines every open cover of X. Therefore $\check{H}^n(X) = H^n(|N(\mathcal{U}_0)|)$.

Example 5.3.3. If $X = \mathbb{S}(S^0)$ is the minimal finite model of S^1, $N(\mathcal{U}_0)$ is a 1-simplex and therefore $\check{H}^1(X) = 0$. On the other hand $H^1(X) = H^1(S^1) = \mathbb{Z}$.

If K is a finite simplicial complex, the cover $\mathcal{U}_0$ of $\mathcal{X}(K)$ satisfies that arbitrary intersections of its elements is empty or homotopically trivial. Indeed, if $\sigma_1, \sigma_2, \ldots, \sigma_r$ are maximal simplices of K, then $\cap U_{\sigma_i}$ is empty or it is $U_{\cap \sigma_i}$. By Theorem 5.3.1, there is a weak homotopy equivalence $|N(\mathcal{U}_0)| \to \mathcal{X}(K)$. Therefore $\check{H}^n(\mathcal{X}(K)) = H^n(|N(\mathcal{U}_0)|) = H^n(\mathcal{X}(K))$, so we have proved

Proposition 5.3.4. *Let* K *be a finite simplicial complex. Then* $\check{H}^n(\mathcal{X}(K)) = H^n(\mathcal{X}(K))$ *for every* $n \geq 0$.

Another proof of the last result can be given invoking a theorem of Dowker [25]. Let V be the set of vertices of K and S the set of its maximal simplices. Define the relation $R \subseteq V \times S$ by $vR\sigma$ if $v \in \sigma$. Dowker considered two simplicial complexes. The simplices of the first complex are the finite subsets of V which are related with a same element of S. This is the original complex K. The simplices of the second complex are the finite subsets of S which are related with a same element of V. This complex is isomorphic to $N(\mathcal{X}(\mathcal{U}_0))$. The Theorem of Dowker concludes that $|K|$ and

$|N(\mathcal{X}(\mathcal{U}_0))|$ are homotopy equivalent. Therefore $H^n(\mathcal{X}(K)) = H^n(|K|) = H^n(|N(\mathcal{X}(\mathcal{U}_0))|) = \check{H}^n(\mathcal{X}(K))$.

Both the Nerve lemma and Dowker's Theorem claim the certain complexes have the same homotopy type. These results can be improved showing that in fact those complexes have the same simple homotopy type. Those versions appear in [6] and follow from a stronger version of Theorem 1.4.19.

We now put the Čech cohomology aside to center our attention in the construction which transforms K in the complex $N(\mathcal{X}(\mathcal{U}_0))$.

Definition 5.3.5. Let K be a finite simplicial complex. The *nerve* of K is the complex $\mathcal{N}(K) = N(\mathcal{X}(\mathcal{U}_0))$. This is the nerve of the cover of $|K|$ given by the maximal simplices. In other words, the vertices of $\mathcal{N}(K)$ are the maximal simplices of K and the simplices of $\mathcal{N}(K)$ are the sets of maximal simplices of K with nonempty intersection. Given $n \geq 2$, we define recursively $\mathcal{N}^n(K) = \mathcal{N}(\mathcal{N}^{n-1}(K))$.

By the arguments above, if K is a finite simplicial complex, $|K|$ and $|\mathcal{N}(K)|$ have the same homotopy type.

Example 5.3.6. Let K be the following simplicial complex

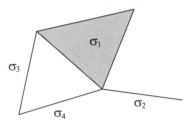

Since K has four maximal simplices, $\mathcal{N}(K)$ has four vertices, and it looks as follows

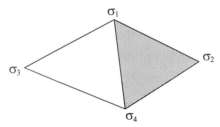

For $n \geq 2$, the complex $\mathcal{N}^n(K)$ is the boundary of a 2-simplex.

If $\mathcal{N}^r(K) = *$ for some $r \geq 1$, then $|K|$ is contractible. But there are contractible complexes such that $\mathcal{N}^r(K)$ is not a point for every r. For instance, if K is the complex of Example 5.1.12, $\mathcal{N}(K)$ has more vertices than K, but $\mathcal{N}^2(K)$ is isomorphic to K. Therefore $\mathcal{N}^r(K) \neq *$ for every r, although $|K|$ is contractible.

We will see that in fact, there is a strong collapse from K to a complex isomorphic to $\mathcal{N}^2(K)$ and that there exists r such that $\mathcal{N}^r(K) = *$ if and only if K is strong collapsible.

Lemma 5.3.7. *Let L be a full subcomplex of a finite simplicial complex K such that every vertex of K which is not in L is dominated by some vertex in L. Then $K \searrow L$.*

Proof. Let v be a vertex of K which is not in L. By hypothesis, v is dominated and then $K \searrow K \smallsetminus v$. Now suppose w is a vertex of $K \smallsetminus v$ which is not in L. Then the link $lk_K(w)$ in K is a simplicial cone aM with $a \in L$. Therefore, the link $lk_{K \smallsetminus v}(w)$ in $K \smallsetminus v$ is $a(M \smallsetminus v)$. By induction $K \smallsetminus v \searrow L$ and then $K \searrow L$. □

Proposition 5.3.8. *Let K be a finite simplicial complex. Then there exists a complex L isomorphic to $\mathcal{N}^2(K)$ such that $K \searrow L$.*

Proof. A vertex of $\mathcal{N}^2(K)$ is a maximal family $\Sigma = \{\sigma_0, \sigma_1, \ldots, \sigma_r\}$ of maximal simplices of K with nonempty intersection. Consider a vertex map $\varphi : \mathcal{N}^2(K) \to K$ such that $\varphi(\Sigma) \in \bigcap_{i=0}^{r} \sigma_i$. This is a simplicial map for if $\Sigma_0, \Sigma_1, \ldots, \Sigma_r$ constitute a simplex of $\mathcal{N}^2(K)$, then there is a common element σ in all of them, which is a maximal simplex of K. Therefore $\varphi(\Sigma_i) \in \sigma$ for every $0 \le i \le r$ and then $\{\varphi(\Sigma_1), \varphi(\Sigma_2), \ldots, \varphi(\Sigma_r)\}$ is a simplex of K.

The vertex map φ is injective. If $\varphi(\Sigma_1) = v = \varphi(\Sigma_2)$ for $\Sigma_1 = \{\sigma_0, \sigma_1, \ldots, \sigma_r\}$, $\Sigma_2 = \{\tau_0, \tau_1, \ldots, \tau_t\}$, then $v \in \sigma_i$ for every $0 \le i \le r$ and $v \in \tau_i$ for every $0 \le i \le t$. Therefore $\Sigma_1 \cup \Sigma_2$ is a family of maximal simplices of K with nonempty intersection. By the maximality of Σ_1 and Σ_2, $\Sigma_1 = \Sigma_1 \cup \Sigma_2 = \Sigma_2$.

Suppose $\Sigma_0, \Sigma_1, \ldots, \Sigma_r$ are vertices of $\mathcal{N}^2(K)$ such that $v_0 = \varphi(\Sigma_0), v_1 = \varphi(\Sigma_1), \ldots, v_r = \varphi(\Sigma_r)$ constitute a simplex of K. Let σ be a maximal simplex of K which contains $v_0, v_1, \ldots, v_r$. Then, by the maximality of the families Σ_i, $\sigma \in \Sigma_i$ for every i and therefore $\{\Sigma_0, \Sigma_1, \ldots, \Sigma_r\}$ is a simplex of $\mathcal{N}^2(K)$.

This proves that $L = \varphi(\mathcal{N}^2(K))$ is a full subcomplex of K which is isomorphic to $\mathcal{N}^2(K)$.

Now, suppose v is a vertex of K which is not in L. Let Σ be the set of maximal simplices of K which contain v. The intersection of the elements of Σ is nonempty, but Σ could be not maximal. Let $\Sigma' \supseteq \Sigma$ be a maximal family of maximal simplices of K with nonempty intersection. Then $v' = \varphi(\Sigma') \in L$ and if σ is a maximal simplex of K which contains v, then $\sigma \in \Sigma \subseteq \Sigma'$. Hence, $v' \in \sigma$. Therefore v is dominated by v'. By Lemma 5.3.7, $K \searrow L$. □

Lemma 5.3.9. *A finite simplicial complex K is minimal if and only if $\mathcal{N}^2(K)$ is isomorphic to K.*

Proof. By Proposition 5.3.8, there exists a complex L isomorphic to $\mathcal{N}^2(K)$ such that $K \searrow L$. Therefore, if K is minimal, $L = K$.

If K is not minimal, there exists a vertex v dominated by other vertex v'. If v is contained in each element of a maximal family Σ of maximal simplices of K with nonempty intersection, then the same occur with v'. Therefore, we can define the map φ of the proof of Proposition 5.3.8 so that v is not in its image. Therefore, $L = \varphi(\mathcal{N}^2(K))$ is isomorphic to $\mathcal{N}^2(K)$ and has less vertices than K. Thus, $\mathcal{N}^2(K)$ and K cannot be isomorphic.   □

The sequence $K, \mathcal{N}^2(K), \mathcal{N}^4(K), \mathcal{N}^6(K), \ldots$ is a decreasing sequence of subcomplexes of K (up to isomorphism). Therefore, there exists $n \geq 1$ such that $\mathcal{N}^{2n}(K)$ and $\mathcal{N}^{2n+2}(K)$ are isomorphic. Then K strongly collapses to a subcomplex L which is isomorphic to $\mathcal{N}^{2n}(K)$ and which is minimal. Thus, we have proved the following

Proposition 5.3.10. *Given a finite simplicial complex K, there exists $n \geq 1$ such that $\mathcal{N}^n(K)$ is isomorphic to the core of K.*

Theorem 5.3.11. *Let K be a finite simplicial complex. Then, K is strong collapsible if and only if there exists $n \geq 1$ such that $\mathcal{N}^n(K)$ is a point.*

Proof. If K is strong collapsible, its core is a point and then, there exists n such that $\mathcal{N}^n(K) = *$ by the previous proposition. If $\mathcal{N}^n(K) = *$ for some n, then $\mathcal{N}^{n+1}(K)$ is also a point. Therefore there exists an even positive integer r such that $\mathcal{N}^r(K) = *$, and $K \searrow\searrow *$ by Proposition 5.3.8.   □

Example 5.3.12. The following complex K is strong collapsible since $\mathcal{N}^3(K) = *$.

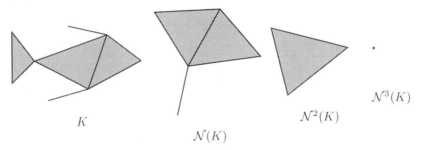

K $\mathcal{N}(K)$ $\mathcal{N}^2(K)$ $\mathcal{N}^3(K)$

Chapter 6
Methods of Reduction

A method of reduction of finite spaces is a technique that allows one to reduce the number of points of a finite topological space preserving some properties of the space.

An important example of a reduction method is described by beat points and was introduced by Stong (see Chap. 1). In that case, the property that is preserved is the homotopy type. Stong's method is effective in the sense that for any finite T_0-space X, one can obtain a space homotopy equivalent to X of minimum cardinality, by applying repeatedly the method of removing beat points.

One of the most important methods of reduction studied in this work is the one described by weak points (see Chap. 4). A removal of a weak point preserves the simple homotopy type. The existence of non-collapsible homotopically trivial finite spaces shows that this method is not an effective way of obtaining minimal finite models. However, a simple homotopy equivalent finite space can always be reached by removing and adding weak points. In some sense this is the best result that can be obtained since there exists no effective reduction method for the weak homotopy type and the simple homotopy type. This is a consequence of the following fact: there is no algorithm which decides whether a finite simplicial complex is contractible or not. This follows in turn from Novikov's result on the undecidability of recognition of spheres (see [49] for instance).

6.1   Osaki's Reduction Methods

Some examples of reduction methods were introduced by Osaki in [65]. In these cases, Osaki presents two methods that allow one to find a quotient of a given finite space such that the quotient map is a weak homotopy equivalence.

J.A. Barmak, *Algebraic Topology of Finite Topological Spaces and Applications*, Lecture Notes in Mathematics 2032, DOI 10.1007/978-3-642-22003-6_6, © Springer-Verlag Berlin Heidelberg 2011

Theorem 6.1.1. *(Osaki) Let X be a finite T_0-space. Suppose there exists $x \in X$ such that $U_x \cap U_y$ is either empty or homotopically trivial for all $y \in X$. Then the quotient map $q : X \to X/U_x$ is a weak homotopy equivalence.*

Proof. Let $y \in X$. If $U_x \cap U_y = \emptyset$, $q^{-1}(U_{qy}) = U_y$. In other case, $q^{-1}(U_{qy}) = U_x \cup U_y$ (see Lemma 2.7.6). In order to apply McCord's Theorem 1.4.2 to the minimal basis of X/U_x, we only have to prove that if $U_x \cap U_y$ is homotopically trivial, then so is $U_x \cup U_y$. If $U_x \cap U_y$ is homotopically trivial, since U_x and U_y are contractible, we obtain from the Mayer-Vietoris sequence for reduced homology that $\widetilde{H}_n(U_x \cup U_y) = 0$ for every $n \geq 0$ and from the Theorem of van Kampen that $U_x \cup U_y$ is simply connected. By Hurewicz's Theorem, it is homotopically trivial. Therefore, Theorem 1.4.2 applies, and q is a weak homotopy equivalence.   □

The process of obtaining X/U_x from X is called an *open reduction*. There is an analogous result for the *minimal closed sets* F_x, i.e. the closures of the one point spaces $\{x\}$. This result follows from the previous one applied to the opposite X^{op}.

Theorem 6.1.2. *(Osaki) Let X be a finite T_0-space. Suppose there exists $x \in X$ such that $F_x \cap F_y$ is either empty or homotopically trivial for all $y \in X$. Then the quotient map $q : X \to X/F_x$ is a weak homotopy equivalence.*

The process of obtaining X/F_x from X is called a *closed reduction*.

Osaki asserts in [65] that he does not know whether by a sequence of reductions, each finite T_0-space can be reduced to the smallest space with the same homotopy groups.

We show with the following example that the answer to this question is negative.

Let $X = \{a_1, b, a_2, c, d, e\}$ be the 6-point T_0-space with the following order: $c, d < a_1$; $c, d, e < b$ and $d, e < a_2$. Let $D_3 = \{c, d, e\}$ be the 3-point discrete space and $Y = \mathbb{S}D_3 = \{a, b, c, d, e\}$ the non-Hausdorff suspension of D_3.

The function $f : X \to Y$ defined by $f(a_1) = f(a_2) = a$, $f(b) = b$, $f(c) = c$, $f(d) = d$ and $f(e) = e$ is continuous because it preserves the order.

In order to prove that f is a weak homotopy equivalence we use the Theorem of McCord 1.4.2. The sets U_y form a basis like cover of Y. It is easy to verify that $f^{-1}(U_y)$ is contractible for each $y \in Y$ and, since U_y is also contractible, the map $f|_{f^{-1}(U_y)} : f^{-1}(U_y) \to U_y$ is a weak homotopy equivalence for each $y \in Y$. Applying Theorem 1.4.2, one proves that f is a weak homotopy equivalence. Therefore X and Y have the same homotopy groups.

Another way to show that X and Y are weak homotopy equivalent is considering the associated polyhedra $|\mathcal{K}(X)|$ and $|\mathcal{K}(Y)|$ which are homotopy equivalent to $S^1 \vee S^1$.

On the other hand, it is easy to see that Osaki's reduction methods cannot be applied to the space X. Therefore his methods are not effective in this case since we cannot obtain, by a sequence of reductions, the smallest space with the same homotopy groups as X.

6.2 γ-Points and One-Point Reduction Methods

In this section we delve deeper into the study of one-point reductions of finite spaces, i.e. reduction methods which consist of removing just one point of the space. Beat points and weak points provide two important examples of one-point reductions. The results of this section are essentially contained in [9].

Recall that $x \in X$ is a weak point if and only if $\hat{C}_x$ is contractible (Remark 4.2.3). This motivates the following definition.

Definition 6.2.1. A point x of a finite T_0-space X is a *γ-point* if $\hat{C}_x$ is homotopically trivial.

Note that weak points are γ-points. It is not difficult to see that both notions coincide in spaces of height less than or equal to 2. This is because any space of height 1 is contractible if and only if it is homotopically trivial. However, this is false for spaces of height greater than 2.

Let x be a γ-point of a finite T_0-space X. We will show that the inclusion $X \smallsetminus \{x\} \hookrightarrow X$ is a weak homotopy equivalence. Note that since $\hat{U}_x$ and $\hat{F}_x$ need not be homotopically trivial, we cannot proceed as we did in Proposition 4.2.4. However, in this case, one has the following pushout

$$
\begin{array}{ccc}
|\mathcal{K}(\hat{C}_x)| & \longrightarrow & |\mathcal{K}(C_x)| \\
\downarrow & & \downarrow \\
|\mathcal{K}(X \smallsetminus \{x\})| & \longrightarrow & |\mathcal{K}(X)|
\end{array}
$$

Where $|\mathcal{K}(\hat{C}_x)| \to |\mathcal{K}(C_x)|$ is a homotopy equivalence and $|\mathcal{K}(\hat{C}_x)| \to |\mathcal{K}(X \smallsetminus \{x\})|$ is a closed cofibration. Therefore $|\mathcal{K}(X \smallsetminus \{x\})| \to |\mathcal{K}(X)|$ is a homotopy equivalence (for more details about this argument see Appendix A.2). This proves the following.

Proposition 6.2.2. *If $x \in X$ is a γ-point, the inclusion $i : X \smallsetminus \{x\} \hookrightarrow X$ is a weak homotopy equivalence.*

This result improves an old result which appears for example in Walker's Thesis [81, Proposition 5.8], which asserts, in the language of finite spaces, that $X \smallsetminus \{x\} \hookrightarrow X$ is a weak homotopy equivalence provided $\hat{U}_x$ or $\hat{F}_x$ is homotopically trivial. By Proposition 6.2.12 below, it is clear that a point x is a γ-point if $\hat{U}_x$ or $\hat{F}_x$ is homotopically trivial, but the converse is false.

We will show that the converse of Proposition 6.2.2 is true in most cases. First, we need some results.

Proposition 6.2.3. *Let x be a point of a finite T_0-space X. The inclusion $i : X \smallsetminus \{x\} \hookrightarrow X$ induces isomorphisms in all homology groups if and only if the subspace $\hat{C}_x$ is acyclic.*

Proof. Apply the Mayer-Vietoris sequence to the triple $(\mathcal{K}(X); \mathcal{K}(C_x), \mathcal{K}(X \smallsetminus \{x\}))$. $\square$

Lemma 6.2.4. *If X and Y are nonempty finite T_0-spaces with n and m connected components respectively, the fundamental group $\pi_1(X \circledast Y)$ is a free product of $(n-1)(m-1)$ copies of $\mathbb{Z}$. In particular if $x \in X$ is neither maximal nor minimal, the fundamental group of $\hat{C}_x = \hat{U}_x \circledast \hat{F}_x$ is a free group.*

Proof. It suffices to show that if K and L are finite simplicial complexes with n and m connected components respectively, then $\pi_1(|K * L|)$ is a free group of rank $(n-1)(m-1)$. Take a vertex v_i in each component K_i of K ($1 \leq i \leq n$) and a vertex w_j in each component L_j of L ($1 \leq j \leq m$). Let M be the full subcomplex of $K * L$ spanned by the the vertices v_i and w_j. Then M is a graph, and an easy computation of its Euler characteristic shows that $\pi_1(|M|)$ is a free group of the desired rank. Let $q : K * L \to M$ be the simplicial map that maps K_i to v_i and L_j to w_j and let $i : M \to K * L$ be the inclusion. Since $qi = 1_M$, $q_* i_* = 1_{E(M,v_1)} : E(M,v_1) \to E(M,v_1)$. It remains to show that $i_* q_* = 1_{E(K*L,v_1)}$, but this follows easily from the next two assertions: any edge-path in $K * L$ with origin and end v_1 is equivalent to an edge-path containing only ordered pairs (u, u') with one vertex in K and the other in L, and an edge-path $(v, w), (w, v')$ with $v, v' \in K$, $w \in L$ is equivalent to $(v, w_j), (w_j, v')$ if $w \in L_j$. $\square$

Theorem 6.2.5. *Let X be a finite T_0-space, and $x \in X$ a point which is neither maximal nor minimal and such that $X \smallsetminus \{x\} \hookrightarrow X$ is a weak homotopy equivalence. Then x is a γ-point.*

Proof. If $X \smallsetminus \{x\} \hookrightarrow X$ is a weak homotopy equivalence, $\hat{C}_x$ is acyclic by Proposition 6.2.3. Then $\pi_1(\hat{C}_x)$ is a perfect group and therefore trivial, by Lemma 6.2.4. Now the result follows from the Hurewicz Theorem. $\square$

The theorem fails if x is maximal or minimal, as the next example shows.

Example 6.2.6. Let X be an acyclic finite T_0-space with nontrivial fundamental group. Let $\mathbb{S}(X) = X \cup \{-1, 1\}$ be its non-Hausdorff suspension. Then $\mathbb{S}(X)$ is also acyclic and $\pi_1(\mathbb{S}(X)) = 0$. Therefore it is homotopically trivial.

Hence, $X \cup \{1\} \hookrightarrow \mathbb{S}(X)$ is a weak homotopy equivalence. However -1 is not a γ-point of $\mathbb{S}(X)$.

An alternative proof of Theorem 6.2.5 without using Lemma 6.2.4 explicitly can be made by arguing that $H_1(\hat{U}_x \circledast \hat{F}_x)$ is a free abelian group of rank $(n-1)(m-1)$ if the spaces $\hat{U}_x$ and $\hat{F}_x$ have n and m connected components. This follows from [57, Lemma 2.1] and implies that either $\hat{U}_x$ or $\hat{F}_x$ is connected. By [57, Lemma 2.2], $\hat{C}_x$ is simply connected.

Using the relativity principle of simple homotopy theory [23, (5.3)] one can prove that if x is a γ-point, $|\mathcal{K}(X \smallsetminus \{x\})| \hookrightarrow |\mathcal{K}(X)|$ is a simple homotopy equivalence. We will see that in fact this holds whenever $X \smallsetminus \{x\} \hookrightarrow X$ is a weak homotopy equivalence.

We will need the following key result ([86, Lemma 10], [23, (20.1)]).

Lemma 6.2.7 (Whitehead). *Let (K, L) be a finite CW-pair such that L is a strong deformation retract of K and such that each connected component of $K \smallsetminus L$ is simply connected. Then $L \hookrightarrow K$ is a simple homotopy equivalence.*

Theorem 6.2.8. *Let X be a finite T_0-space and let $x \in X$. If the inclusion $i : X \smallsetminus \{x\} \hookrightarrow X$ is a weak homotopy equivalence, it induces a simple homotopy equivalence $|\mathcal{K}(X \smallsetminus \{x\})| \to |\mathcal{K}(X)|$. In particular $X \smallsetminus \{x\} \diagup\!\!\!\diagdown X$.*

Proof. Since $|\mathcal{K}(X \smallsetminus \{x\})|$ is a strong deformation retract of $|\mathcal{K}(X)|$ and the open star of x,

$$\overset{\circ}{st}(x) = |\mathcal{K}(X)| \smallsetminus |\mathcal{K}(X \smallsetminus \{x\})|$$

is contractible, then by Lemma 6.2.7, $|\mathcal{K}(X \smallsetminus \{x\})| \hookrightarrow |\mathcal{K}(X)|$ is a simple homotopy equivalence. $\qquad\square$

This result essentially shows that one-point reductions are not sufficient to describe all weak homotopy types of finite spaces. They are sufficient, though, to reach all finite models of spaces with trivial Whitehead group by Corollary 4.2.13.

On the other hand, note that the fact that $X \smallsetminus \{x\}$ and X have the same weak homotopy type does not imply that the inclusion $X \smallsetminus \{x\} \hookrightarrow X$ is a weak homotopy equivalence.

Definition 6.2.9. *If $x \in X$ is a γ-point, we say that there is an* elementary γ-collapse *from X to $X \smallsetminus \{x\}$. A finite T_0-space X* γ-collapses *to Y if there is a sequence of elementary γ-collapses that starts in X and ends in Y. We denote this by $X \searrow\!\!\!\!\!_{\gamma}\, Y$. If X γ-collapses to a point, we say that it is* γ-collapsible.

In contrast to collapses, a γ-collapse does not induce in general a collapse between the associated simplicial complexes. For example, if K is any triangulation of the Dunce Hat (see Fig. 4.5 in page 60), $\mathbb{C}(\mathcal{X}(K)) \searrow\!\!\!\!\!_{\gamma}\, \mathcal{X}(K)$, but $aK' \searrow\!\!\!\!\!_{\gamma} K'$ since K' is not collapsible (see Lemma 4.1.1). However, if $X \searrow\!\!\!\!\!_{\gamma} Y$, then $X \diagup\!\!\!\diagdown Y$ by Theorem 6.2.8 and then $\mathcal{K}(X)$ has the same simple homotopy type as $\mathcal{K}(Y)$.

Recall that $f : X \to Y$ is said to be distinguished if $f^{-1}(U_y)$ is contractible for every $y \in Y$. Distinguished maps are simple homotopy equivalences (see Sect. 4.4). The following result generalizes that fact.

Proposition 6.2.10. *Let $f : X \to Y$ be a map between finite T_0-spaces such that $f^{-1}(U_y)$ is homotopically trivial for every $y \in Y$. Then f is a simple homotopy equivalence.*

Proof. Consider the non-Hausdorff mapping cylinder $B(f)$ with the inclusions $i : X \hookrightarrow B(f)$, $j : Y \hookrightarrow B(f)$. Using the same proof of Lemma 4.2.7, one can show that $B(f) \searrow X$, while $B(f) \searrow Y$ by Lemma 2.8.2. Then i and j are simple homotopy equivalences by Theorem 6.2.8, and since $jf \simeq i$, so is f. $\square$

Note that in the hypothesis of the last proposition, every space Z with $f(X) \subseteq Z \subseteq Y$ has the simple homotopy type of Y, because in this case $f : X \to Z$ also satisfies the hypothesis of above.

Remark 6.2.11. The quotient maps of Theorems 6.1.1 and 6.1.2 are simple homotopy equivalences.

We finish this section analyzing the relationship between γ-collapsibility and joins.

Proposition 6.2.12. *Let X and Y be finite T_0-spaces. Then*

(i) $X \circledast Y$ is homotopically trivial if X or Y is homotopically trivial.
(ii) $X \circledast Y$ is γ-collapsible if X or Y is γ-collapsible.

Proof. If X or Y is homotopically trivial, $|\mathcal{K}(X)|$ or $|\mathcal{K}(Y)|$ is contractible and then so is $|\mathcal{K}(X)| * |\mathcal{K}(Y)| = |\mathcal{K}(X \circledast Y)|$. Therefore $X \circledast Y$ is homotopically trivial.

The proof of (ii) follows as in Proposition 2.7.3. If $x_i \in X_i$ is a γ-point, $\hat{C}_{x_i}^{X_i \circledast Y} = \hat{C}_{x_i}^{X_i} \circledast Y$ is homotopically trivial by item (i) and then x_i is a γ-point of $X_i \circledast Y$. $\square$

There is an analogous result for acyclic spaces that follows from the Künneth formula for joins [57].

Note that the converse of these results are false. To see this, consider two finite simply connected simplicial complexes K, L such that $H_2(|K|) = \mathbb{Z}_2$ is the cyclic group of two elements, $H_2(|L|) = \mathbb{Z}_3$ is the cyclic group of three elements and $H_n(|K|) = H_n(|L|) = 0$ for every $n \geq 3$. Then $\mathcal{X}(K)$ and $\mathcal{X}(L)$ are not acyclic, but $\mathcal{X}(K) \circledast \mathcal{X}(L)$, which is weak homotopy equivalent to $|K| * |L|$, is acyclic by the Künneth formula and, since it is simply connected (see [57] or Lemma 6.2.4), it is homotopically trivial.

A counterexample for the converse of item (ii) is the following.

Example 6.2.13. Let K be a triangulation of the Dunce Hat. Then $\mathcal{X}(K)$ is a homotopically trivial finite space of height 2. The non-Hausdorff suspension

$\mathbb{S}(\mathcal{X}(K)) = \mathcal{X}(K) \circledast S^0 = \mathcal{X}(K) \cup \{-1, 1\}$ is γ-collapsible. The point 1 is a γ-point of $\mathbb{S}(\mathcal{X}(K))$ since $\hat{C}_1 = \mathcal{X}(K)$ is homotopically trivial. The subspace $\mathbb{S}(\mathcal{X}(K)) \smallsetminus \{1\}$ has maximum and in particular it is contractible, and therefore γ-collapsible. However $\mathcal{X}(K)$ is not collapsible since K' is not collapsible, and then $\mathbb{S}(\mathcal{X}(K))$ is not collapsible by Proposition 4.3.4. Furthermore, $\mathcal{X}(K)$ and S^0 are not γ-collapsible either because they are non-collapsible spaces of height less than or equal to 2. Therefore $\mathbb{S}(\mathcal{X}(K))$ is a γ-collapsible space which is the join of two non γ-collapsible spaces. Moreover, it is a γ-collapsible space which is not collapsible.

Chapter 7
h-Regular Complexes and Quotients

The results of McCord show that each compact polyhedron $|K|$ can be modeled, up to weak homotopy, by a finite space $\mathcal{X}(K)$. It is not hard to prove that this result can be extended to the so called *regular* CW-complexes. In this chapter we introduce a new class of complexes, generalizing the notion of simplicial complex and of regular complex, and we prove that they also can be modeled by their face posets. This can be used to find smaller models of well-known spaces. The relationship with collapsibility is also studied. The ideas developed in the first section are then used to obtain an exact sequence of homology groups for finite spaces.

The results of this chapter are partially contained in [9]. This is probably the most technical part of the book and requires some familiarity with the theory of CW-complexes. The reader who is not an expert in CW-complexes is invited to consult [28, 38]. Some basic definitions and properties can be found in Appendix A.2.

7.1 h-Regular CW-Complexes and Their Associated Finite Spaces

A CW-complex K is said to be *regular* if for each (open) cell e^n, the characteristic map $D^n \to \overline{e^n}$ is a homeomorphism, or equivalently, the attaching map $S^{n-1} \to K$ is a homeomorphism onto its image $\dot{e}^n$, the boundary of e^n. In this case, it can be proved that the closure $\overline{e^n}$ of each cell is a subcomplex, which is equivalent to saying that $\dot{e}^n$ is a subcomplex (see [28, Theorem 1.4.10]).

A cell e of a regular complex K is a *face* of a cell e' if $e \subseteq \overline{e'}$. This will be denoted by $e \le e'$. The *barycentric subdivision* K' is the simplicial complex whose vertices are the cells of K and whose simplices are the

J.A. Barmak, *Algebraic Topology of Finite Topological Spaces and Applications*, Lecture Notes in Mathematics 2032,
DOI 10.1007/978-3-642-22003-6_7, © Springer-Verlag Berlin Heidelberg 2011

sets $\{e_1, e_2, \ldots, e_n\}$ such that e_i is a face of e_{i+1}. The polyhedron $|K'|$ is homeomorphic to K (see [45, Theorem 1.7] for instance).

We can define, as in the case of simplicial complexes, the face poset $\mathcal{X}(K)$ of a regular complex K, which is the set of cells ordered by $\leq$. Note that $\mathcal{K}(\mathcal{X}(K)) = K'$ and therefore $\mathcal{X}(K)$ is a finite model of K, i.e. it has the same weak homotopy type as K.

Example 7.1.1. The following figure (Fig. 7.1) shows a regular structure for the real projective plane $\mathbb{R}P^2$. The edges are identified in the way indicated by the arrows. It has three 0-cells, six 1-cells and four 3-cells. Therefore its face poset has 13 points (Fig. 7.2). It is unknown to the author whether this is a minimal finite model of $\mathbb{R}P^2$. This finite space appears also in [36, Proposition 4.1] obtained by a different method.

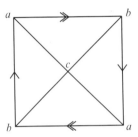

Fig. 7.1 $\mathbb{R}P^2$

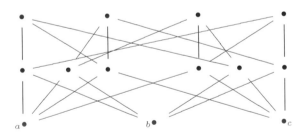

Fig. 7.2 A finite model of $\mathbb{R}P^2$

We introduce now the concept of *h-regular complex*, generalizing the notion of regular complex. Given an h-regular complex K, one can define $\mathcal{X}(K)$ as before. In general, K and $\mathcal{K}(\mathcal{X}(K))$ are not homeomorphic. However we prove that $\mathcal{X}(K)$ is a finite model of K.

Definition 7.1.2. A CW-complex K is *h-regular* if the attaching map of each cell is a homotopy equivalence with its image and the closed cells $\overline{e^n}$ are subcomplexes of K.

In particular, regular complexes are h-regular.

Proposition 7.1.3. *Let $K = L \cup e^n$ be a CW-complex such that $\dot{e}^n$ is a subcomplex of L. Then $\overline{e^n}$ is contractible if and only if the attaching map $\varphi : S^{n-1} \to \dot{e}^n$ of the cell e^n is a homotopy equivalence.*

Proof. Suppose $\varphi : S^{n-1} \to \dot{e}^n$ is a homotopy equivalence. Since $S^{n-1} \hookrightarrow D^n$ is a closed cofibration, the characteristic map $\psi : D^n \to \overline{e^n}$ is also a homotopy equivalence by the gluing theorem A.2.5.

Suppose now that $\overline{e^n}$ is contractible. The map $\overline{\psi} : D^n/S^{n-1} \to \overline{e^n}/\dot{e}^n$ is a homeomorphism and therefore it induces isomorphisms in homology and, since $\overline{e^n}$ is contractible, by the long exact sequence of homology it follows that $\varphi_* : H_k(S^{n-1}) \to H_k(\dot{e}^n)$ is an isomorphism for every k.

If $n \geq 3$, $\pi_1(\dot{e}^n) = \pi_1(\overline{e^n}) = 0$ and by a theorem of Whitehead ([38, Corollary 4.33]), φ is a homotopy equivalence. If $n = 2$, $\dot{e}^n$ is just a graph and since $\varphi_* : H_1(S^1) \to H_1(\dot{e}^n)$ is an isomorphism, the attaching map φ is a homotopy equivalence. Finally, if $n = 1$, since the cell is contractible, φ is one-to-one and therefore a homeomorphism.   □

Corollary 7.1.4. *A CW-complex is h-regular if and only if the closed cells are contractible subcomplexes.*

Example 7.1.5. The following are four different h-regular structures for the Dunce Hat (Fig. 4.5) which are not regular structures. In each example the edges are identified in the way indicated by the arrows.

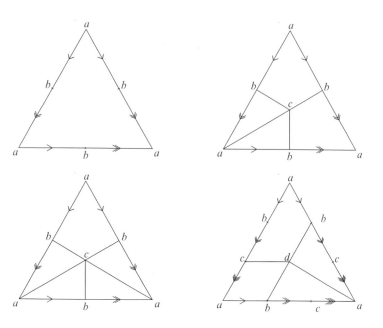

For an h-regular complex K, we also define the *associated finite space* (or *face poset*) $\mathcal{X}(K)$ as the poset of cells of K ordered by the face relation $\le$, as in the regular case. Note that since closed cells are subcomplexes, $e \le e'$ if and only if $\bar{e} \subseteq \overline{e'}$.

The proof of the following lemma is standard.

Lemma 7.1.6. *Let $K \cup e$ be a CW-complex, let $\psi : D^n \to \bar{e}$ be the characteristic map of the cell e and let A be a subspace of $\dot{e}$. We denote $C_e(A) = \{\psi(x) \mid x \in D^n \smallsetminus \{0\}, \ \psi(\frac{x}{\|x\|}) \in A\} \subseteq \bar{e}$. Then*

1. *If $A \subseteq \dot{e}$ is open, $C_e(A) \subseteq \bar{e}$ is open.*
2. *$A \subseteq C_e(A)$ is a strong deformation retract.*

Theorem 7.1.7. *If K is a finite h-regular complex, $\mathcal{X}(K)$ is a finite model of K.*

Proof. We define recursively a weak homotopy equivalence $f_K : K \to \mathcal{X}(K)$.

Assume $f_{K^{n-1}} : K^{n-1} \to \mathcal{X}(K^{n-1}) \subseteq \mathcal{X}(K)$ is already defined and let $x = \psi(a)$ be a point in an n-cell e^n with characteristic map $\psi : D^n \to \overline{e^n}$. If $a = 0 \in D^n$, define $f_K(x) = e^n$. Otherwise, define $f_K(x) = f_{K^{n-1}}(\psi(\frac{a}{\|a\|}))$.

In particular note that if $e^0 \in K$ is a 0-cell, $f_K(e^0) = e^0 \in \mathcal{X}(K)$. Notice also that if L is a subcomplex of K, $f_L = f_K|_L$.

We will show by induction on the number of cells of K, that for every cell $e \in K$, $f_K^{-1}(U_e)$ is open and contractible. This will prove that f_K is continuous and, by McCord's Theorem 1.4.2, a weak homotopy equivalence.

Let e be a cell of K. Suppose first that there exists a cell of K which is not contained in $\bar{e}$. Take a maximal cell e' (with respect to the face relation $\le$) with this property. Then $L = K \smallsetminus e'$ is a subcomplex and by induction, $f_L^{-1}(U_e)$ is open in L. It follows that $f_L^{-1}(U_e) \cap \dot{e}' \subseteq \dot{e}'$ is open and by the previous lemma, $C_{e'}(f_L^{-1}(U_e) \cap \dot{e}') \subseteq \overline{e'}$ is open. Therefore,

$$f_K^{-1}(U_e) = f_L^{-1}(U_e) \cup C_{e'}(f_L^{-1}(U_e) \cap \dot{e}')$$

is open in K.

Moreover, since $f_L^{-1}(U_e) \cap \dot{e}' \subseteq C_{e'}(f_L^{-1}(U_e) \cap \dot{e}')$ is a strong deformation retract, so is $f_L^{-1}(U_e) \subseteq f_K^{-1}(U_e)$. By induction, $f_K^{-1}(U_e)$ is contractible.

In the case that every cell of K is contained in $\bar{e}$, $f_K^{-1}(U_e) = \bar{e} = K$, which is open and contractible.   □

As an application we deduce that the finite spaces associated to the h-regular structures of the Dunce Hat considered in Example 7.1.5 are all homotopically trivial. The first one is a contractible space of 5 points, the second one is a collapsible and non-contractible space of 13 points and the last two are non-collapsible spaces of 15 points since they do not have weak

points. In Fig. 7.3 we exhibit the Hasse diagram of the space associated to
the third h-regular structure of the Dunce Hat.

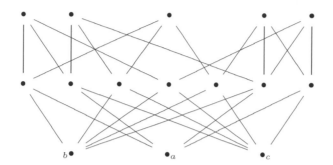

Fig. 7.3 A homotopically trivial non-collapsible space of 15 points

Example 7.1.8. Let K be the space which is obtained from a square by
identifying all its edges as indicated.

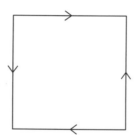

We verify that K is homotopy equivalent to S^2 using techniques of finite
spaces. Consider the following h-regular structure of K

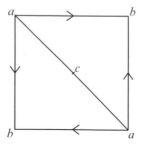

which consists of three 0-cells, three 1-cells and two 2-cells. The Hasse
diagram of the associated finite space $\mathcal{X}(K)$ is

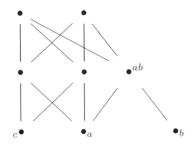

The 0-cell b is an up beat point of $\mathcal{X}(K)$ and the 1-cell ab is a down beat point
of $\mathcal{X}(K) \smallsetminus \{b\}$. Therefore K is weak homotopy equivalent to $\mathcal{X}(K) \smallsetminus \{b, ab\}$
which is a (minimal) finite model of S^2 (see Chap. 3). In fact $\mathcal{X}(K) \smallsetminus \{b, ab\} =$
$S^0 \circledast S^0 \circledast S^0$ is weak homotopy equivalent to $S^0 * S^0 * S^0 = S^2$.

In [14, Proposition 3.1], Björner gives a characterization of the posets
which are face posets of finite regular CW-complexes. They are the finite
T_0-spaces X such that $|\mathcal{K}(\hat{U}_x)|$ is homeomorphic to a sphere for every $x \in X$.
The analogous result for h-regular complexes is an open problem. It is easy
to prove that if X is the finite space associated to a finite h-regular complex,
then $\hat{U}_x$ is a finite model of a sphere of dimension $ht(x)-1 = ht(\hat{U}_x)$. However
it is unknown whether this is a sufficient condition for being the face poset
of an h-regular complex.

In Chap. 4 we proved that a collapse $K \searrow L$ of finite simplicial complexes
induces a collapse $\mathcal{X}(K) \searrow \mathcal{X}(L)$ between the associated finite spaces. This
is not true when K and L are regular complexes. Consider $L = \mathcal{K}(W)$ the
simplicial complex associated to the Wallet W (see Fig. 4.2 in page 54), and
K the CW-complex obtained from L by attaching a regular 2-cell e^2 with
boundary $\mathcal{K}(\{a, b, c, d\})$ and a regular 3-cell e^3 with boundary $L \cup e^2$.

Note that the complex K is regular and collapses to L, but $\mathcal{X}(K) =$
$\mathcal{X}(L) \cup \{e^2, e^3\}$ does not collapse to $\mathcal{X}(L)$ because $\hat{U}_{e^3}^{\mathcal{X}(K) \smallsetminus \{e^2\}} = \mathcal{X}(L) = W'$
is not contractible. However, one can prove that a collapse $K \searrow L$ between
h-regular CW-complexes induces a γ-collapse $\mathcal{X}(K) \searrow^\gamma \mathcal{X}(L)$.

Theorem 7.1.9. *Let L be a subcomplex of an h-regular complex K. If $K \searrow L$,
then $\mathcal{X}(K) \searrow^\gamma \mathcal{X}(L)$.*

Proof. Assume $K = L \cup e^n \cup e^{n+1}$. Then e^n is an up beat point of $\mathcal{X}(K)$.
Since $K \searrow L$, $\overline{e^{n+1}} \searrow L \cap \overline{e^{n+1}} = \dot{e}^{n+1} \smallsetminus e^n$. In particular $\dot{e}^{n+1} \smallsetminus e^n$ is
contractible and then

$$\hat{C}_{e^{n+1}}^{\mathcal{X}(K) \smallsetminus \{e^n\}} = \mathcal{X}(\dot{e}^{n+1} \smallsetminus e^n)$$

is homotopically trivial. Therefore

$$\mathcal{X}(K) \searrow^{e} \mathcal{X}(K) \smallsetminus \{e^n\} \searrow \mathcal{X}(L).$$

□

We study the relationship between the weak homotopy equivalence $f_K :$ $|K| \to \mathcal{X}(K)$ defined in Theorem 7.1.7 and the McCord map $\mu_K :$ $|K| \to \mathcal{X}(K)$. We will prove that both maps coincide if we take convenient characteristic maps for the cells of the polyhedron $|K|$.

Let σ be an n-simplex of the simplicial complex K. Let $\varphi : S^{n-1} \to \dot\sigma$ be a homeomorphism. Define the characteristic map $\overline{\varphi} : D^n \to \overline{\sigma}$ of the cell σ by

$$\overline{\varphi}(x) = \begin{cases} (1- \parallel x \parallel)b(\sigma)+ \parallel x \parallel \varphi(\frac{x}{\parallel x \parallel}) & \text{if } x \neq 0 \\ b(\sigma) & \text{if } x = 0 \end{cases}$$

Here $b(\sigma) \in \overline{\sigma}$ denotes the barycenter of σ. Clearly $\overline{\varphi}$ is continuous and bijective and therefore a homeomorphism.

Definition 7.1.10. We say that the polyhedron $|K|$ has a *convenient cell structure* (as a CW-complex) if the characteristic maps of the cells are defined as above.

Proposition 7.1.11. *Let K be a finite simplicial complex. Consider a convenient cell structure for $|K|$. Then f_K and μ_K coincide.*

Proof. Let $x \in |K|$, contained in an open n-simplex σ. Let $\varphi : S^{n-1} \to |K|$ be the attaching map of the cell σ, and $\overline{\varphi} : D^n \to \overline{\sigma}$ its characteristic map. If x is the barycenter of σ, $f_K(x) = f_K(\overline{\varphi}(0)) = \sigma \in \mathcal{X}(K)$ and $\mu_K(x) = \mu_{\mathcal{X}(K)}s_K^{-1}(b(\sigma)) = \mu_{\mathcal{X}(K)}(\sigma) = \sigma$. Assume then that $x = \overline{\varphi}(y)$ with $y \neq 0$. Thus, $f_K(x) = f_K(\varphi(\frac{y}{\parallel y \parallel}))$. Then, by an inductive argument,

$$f_K(x) = \mu_K(\varphi(\frac{y}{\parallel y \parallel})) = \mu_{\mathcal{X}(K)}(s_K^{-1}\varphi(\frac{y}{\parallel y \parallel})).$$

On the other hand,

$$\mu_K(x) = \mu_{\mathcal{X}(K)}s_K^{-1}(\overline{\varphi}(y)) = \mu_{\mathcal{X}(K)}s_K^{-1}((1- \parallel y \parallel)b(\sigma)+ \parallel y \parallel \varphi(\frac{y}{\parallel y \parallel}))$$

$$= \mu_{\mathcal{X}(K)}((1- \parallel y \parallel)\sigma+ \parallel y \parallel s_K^{-1}\varphi(\frac{y}{\parallel y \parallel})).$$

Finally, $s_K^{-1}\varphi(\frac{y}{\parallel y \parallel}) \in |(\dot\sigma)'|$ and then,

$$\mu_{\mathcal{X}(K)}((1- \parallel y \parallel)\sigma+ \parallel y \parallel s_K^{-1}\varphi(\frac{y}{\parallel y \parallel}))$$

$$= \min(support((1- \parallel y \parallel)\sigma+ \parallel y \parallel s_K^{-1}\varphi(\frac{y}{\parallel y \parallel})))$$

$$= \min(\{\sigma\} \cup support(s_K^{-1}\varphi(\frac{y}{\| y \|})))$$

$$= \min(support(s_K^{-1}\varphi(\frac{y}{\| y \|}))) = \mu_{\mathcal{X}(K)}(s_K^{-1}\varphi(\frac{y}{\| y \|})).$$

Thus, $f_K(x) = \mu_K(X)$.   □

7.2   Quotients of Finite Spaces: An Exact Sequence for Homology Groups

For CW-pairs (Z, W), there exists a long exact sequence of reduced homology groups

$$\cdots \longrightarrow \tilde{H}_n(W) \longrightarrow \tilde{H}_n(Z) \longrightarrow \tilde{H}_n(Z/W) \longrightarrow \tilde{H}_{n-1}(W) \longrightarrow \cdots$$

More generally, this holds for any good pair (Z, W); i.e. a topological pair such that W is closed in Z and is a deformation retract of some neighborhood in Z [38, Theorem 2.13]. When W is an arbitrary subspace of a finite space Z, one does not have such a sequence in general. For a pair of finite spaces (X, A), $H_n(X, A)$ and $\tilde{H}_n(X/A)$ need not be isomorphic (see Example 2.7.9). However, we will prove that if A is a subspace of a finite T_0-space X, there is a long exact sequence

$$\cdots \longrightarrow \tilde{H}_n(A') \longrightarrow \tilde{H}_n(X') \longrightarrow \tilde{H}_n(X'/A') \longrightarrow \tilde{H}_{n-1}(A') \longrightarrow \cdots$$

of the reduced homology groups of the subdivisions of X and A and their quotient. In fact, in this case we will prove that $\tilde{H}_n(X'/A') = H_n(X, A) = H_n(X', A')$.

Recall that if W is a subcomplex of a CW-complex Z, Z/W is CW-complex with one n-cell for every n-cell of Z which is not a cell of W and an extra 0-cell. The n-squeleton $(Z/W)^n$ is the quotient Z^n/W^n. If $\overline{e^n}$ is a closed n-cell of Z which is not in W, there is a corresponding closed n-cell $q(\overline{e^n})$ in Z/W where $q : Z \to Z/W$ is the quotient map. If $\varphi : S^{n-1} \to Z^{n-1}$ is the attaching map of e^n and $\overline{\varphi} : D^n \to \overline{e^n}$ its characteristic map, $q\varphi : S^{n-1} \to Z^{n-1}/W^{n-1}$ and $q\overline{\varphi} : D^n \to q(e^n)$ are respectively, the attaching and characteristic maps of the corresponding cell $\widetilde{e}^n$ in Z/W.

Theorem 7.2.1. *Let K be a finite simplicial complex and let $L \subseteq K$ be a full subcomplex. Then $|K|/|L|$ is an h-regular CW-complex and $\mathcal{X}(|K|/|L|)$ is homeomorphic to $\mathcal{X}(K)/\mathcal{X}(L)$.*

Proof. Let σ be an n-simplex of K which is not a simplex of L. If σ intersects L, then $\sigma \cap L = \tau$ is a proper face of σ. In particular $\overline{\tau}$ is contractible and therefore the corresponding closed cell $q(\overline{\sigma}) = \overline{\sigma}/\overline{\tau} \subseteq |K|/|L|$ is homotopy equivalent to $\overline{\sigma}$ which is contractible (see Proposition A.2.7). Thus, closed cells of $|K|/|L|$ are contractible subcomplexes. By Corollary 7.1.4, $|K|/|L|$ is h-regular.

Now, if τ and σ are simplices of K which are not in L, then $\widetilde{\tau} \leq \widetilde{\sigma}$ in $\mathcal{X}(|K|/|L|)$ if and only if $q(\overline{\tau}) = \widetilde{\widetilde{\tau}} \subseteq \widetilde{\widetilde{\sigma}} = q(\overline{\sigma})$ if and only if τ is a face of σ in K if and only if $\tau \leq \sigma$ in $\mathcal{X}(K)/\mathcal{X}(L)$. Finally, if $\tau \in L$ and $\sigma \notin L$, $\widetilde{\tau} < \widetilde{\sigma}$ in $\mathcal{X}(|K|/|L|)$ if and only if $q(\overline{\tau}) \subset q(\overline{\sigma})$ if and only if $\sigma \cap L \neq \emptyset$ if and only if $\tau < \sigma$ in $\mathcal{X}(K)/\mathcal{X}(L)$. Therefore, $\mathcal{X}(|K|/|L|)$ and $\mathcal{X}(K)/\mathcal{X}(L)$ are homeomorphic. $\square$

Corollary 7.2.2. *Let X be a finite T_0-space and $A \subseteq X$ a subspace. Then $\mathcal{X}(|\mathcal{K}(X)|/|\mathcal{K}(A)|)$ is homeomorphic to X'/A'. In particular $|\mathcal{K}(X)|/|\mathcal{K}(A)|$ and $|\mathcal{K}(X'/A')|$ are homotopy equivalent.*

Proof. Apply Theorem 7.2.1 to $K = \mathcal{K}(X)$ and the full subcomplex $L = \mathcal{K}(A)$. $\square$

Corollary 7.2.3. *If A is a subspace of a finite T_0-space X, $H_n(X, A) = \widetilde{H}_n(X'/A')$ for every $n \geq 0$.*

Proof. By the naturality of the long exact sequence of homology, the McCord map $\mu_X : |\mathcal{K}(X)| \to X$ induces isomorphisms $H_n(|\mathcal{K}(X)|, |\mathcal{K}(A)|) \to H_n(X, A)$. Thus, $H_n(X, A) = H_n(|\mathcal{K}(X)|, |\mathcal{K}(A)|) = \widetilde{H}_n(|\mathcal{K}(X)|/|\mathcal{K}(A)|) = \widetilde{H}_n(|\mathcal{K}(X'/A')|) = \widetilde{H}_n(X'/A')$. $\square$

Example 2.7.9 shows that $H_n(X, A)$ is not isomorphic to $\widetilde{H}_n(X/A)$ in general.

Proposition 7.2.4. *Let L be a full subcomplex of a finite simplicial complex K. Let $f_K : |K| \to \mathcal{X}(K)$, $f_{K/L} : |K|/|L| \to \mathcal{X}(|K|/|L|)$ be the weak homotopy equivalences constructed in Theorem 7.1.7 (for some characteristic maps of the cells of $|K|$). Let $q : |K| \to |K|/|L|$ and $\widetilde{q} : \mathcal{X}(K) \to \mathcal{X}(K)/\mathcal{X}(L)$ be the quotient maps and let $h : \mathcal{X}(|K|/|L|) \to \mathcal{X}(K)/\mathcal{X}(L)$ be the homeomorphism defined by $h(\widetilde{\sigma}) = \widetilde{q}(\sigma)$. Then the following diagram commutes*

$$
\begin{array}{ccc}
|K| & \xrightarrow{\ q\ } & |K|/|L| \\
\Big\downarrow{\scriptstyle f_K} & & \Big\downarrow{\scriptstyle hf_{K/L}} \\
\mathcal{X}(K) & \xrightarrow{\ \widetilde{q}\ } & \mathcal{X}(K)/\mathcal{X}(L).
\end{array}
$$

Proof. Let $x \in |K|$, $x \in e^n$, an open n-simplex. We prove that $\widetilde{q} f_K(x) = h f_{K/L} q(x)$ by induction in n. Note that this is clear if $x \in |L|$, so we suppose $x \notin |L|$. If $n = 0$, $h f_{K/L} q(e^0) = h f_{K/L}(\widetilde{e}^0) = h(\widetilde{e}^0) = \widetilde{q}(e^0) = \widetilde{q} f_K(e^0)$. Assume then that $n > 0$, $x \in e^n$. Let $\varphi : S^{n-1} \to |K|$ and $\overline{\varphi} : D^n \to \overline{e^n}$ be the attaching and characteristic maps of e^n. Since e^n is not a simplex of L, $\overline{e}^n$ is a cell of $|K|/|L|$ with attaching map $q\varphi : S^{n-1} \to |K|/|L|$ and characteristic map $q\overline{\varphi} : D^n \to q(e^n)$. Let y in the interior of the disk D^n such that $x = \overline{\varphi}(y)$. By definition of $f_{K/L}$,

$$f_{K/L}(q(x)) = f_{K/L}((q\overline{\varphi})(y))) = \begin{cases} f_{K/L}((q\varphi)(\frac{y}{\|y\|})) & \text{if } y \neq 0 \\ \overline{e}^n & \text{if } y = 0 \end{cases}$$

If $y \neq 0$, $h f_{K/L}(q(x)) = h f_{K/L} q(\varphi(\frac{y}{\|y\|})) = \widetilde{q} f_K(\varphi(\frac{y}{\|y\|})) = \widetilde{q} f_K(x)$ by induction. If $y = 0$, $h f_{K/L}(x) = h(\widetilde{e}^n) = \widetilde{q}(e^n) = \widetilde{q} f_K(x)$. This proves that $\widetilde{q} f_K(x) = h f_{K/L} q(x)$. $\square$

Let $\partial : \widetilde{H}_n(|K|/|L|) \to \widetilde{H}_{n-1}(|L|)$ be the connecting homomorphism of the long exact sequence of reduced homology. Define $\widetilde{\partial} = f_{L*} \partial ((h f_{K/L})_*)^{-1} : \widetilde{H}_n(\mathcal{X}(K)/\mathcal{X}(L)) \to \widetilde{H}_n(\mathcal{X}(L))$. By the previous results, there exists a long exact sequence

$$\to \widetilde{H}_n(\mathcal{X}(L)) \xrightarrow{i_*} \widetilde{H}_n(\mathcal{X}(K)) \xrightarrow{\widetilde{q}_*} \widetilde{H}_n(\mathcal{X}(K)/\mathcal{X}(L)) \xrightarrow{\widetilde{\partial}} \widetilde{H}_{n-1}(\mathcal{X}(L)) \to$$

$$(7.1)$$

Corollary 7.2.5. *Let A be a subspace of a finite T_0-space X. There exists a long exact sequence*

$$\cdots \longrightarrow \widetilde{H}_n(A') \xrightarrow{i_*} \widetilde{H}_n(X') \xrightarrow{\widetilde{q}_*} \widetilde{H}_n(X'/A') \xrightarrow{\widetilde{\partial}} \widetilde{H}_{n-1}(A') \longrightarrow \cdots$$

$$(7.2)$$

which is natural in the following sense: if $g : (X, A) \to (Y, B)$ is a map of pairs, there is a commutative diagram

$$\cdots \longrightarrow \widetilde{H}_n(A') \xrightarrow{i_*} \widetilde{H}_n(X') \xrightarrow{\widetilde{q}_*} \widetilde{H}_n(X'/A') \xrightarrow{\widetilde{\partial}} \widetilde{H}_{n-1}(A') \longrightarrow \cdots$$
$$\downarrow g'_* \qquad\qquad \downarrow g'_* \qquad\qquad \downarrow \overline{g'}_* \qquad\qquad \downarrow g'_*$$
$$\cdots \longrightarrow \widetilde{H}_n(B') \xrightarrow{i_*} \widetilde{H}_n(Y') \xrightarrow{\widetilde{q}_*} \widetilde{H}_n(Y'/B') \xrightarrow{\widetilde{\partial}} \widetilde{H}_{n-1}(B') \longrightarrow \cdots$$

$$(7.3)$$

where $g' = \mathcal{X}(\mathcal{K}(g))$ is the induced map in the subdivisions.

Proof. Consider a convenient cell structure for $|\mathcal{K}(X)|$. Taking $K = \mathcal{K}(X)$ and $L = \mathcal{K}(A)$ in (7.1) one obtains the long exact sequence (7.2) with the connecting morphism $\tilde{\partial}$ defined as above for the maps f_K and $f_{K/L}$ induced by the cell structure of $|\mathcal{K}(X)|$.

The first two squares of (7.3) commute before taking homology. We only have to prove the commutativity of the third square.

Consider the following cube,

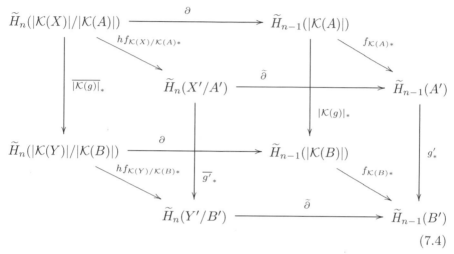

$$(7.4)$$

The top and bottom faces of the cube commute by definition of $\tilde{\partial}$. The back face commute by the naturality of the long exact sequence for CW-complexes. Therefore, to prove that the front face commutes, we only have to check that the left and right faces do. To achieve this, we prove that these two squares commute up to homotopy:

$$
\begin{array}{ccc}
|\mathcal{K}(A)| & \xrightarrow{f_{\mathcal{K}(A)}} & A' \\
\downarrow{\scriptstyle |\mathcal{K}(g)|} & & \downarrow{\scriptstyle g'} \\
|\mathcal{K}(B)| & \xrightarrow{f_{\mathcal{K}(B)}} & B'
\end{array}
\qquad
\begin{array}{ccc}
|\mathcal{K}(X)|/|\mathcal{K}(A)| & \xrightarrow{hf_{\mathcal{K}(X)/\mathcal{K}(A)}} & X'/A' \\
\downarrow{\scriptstyle \overline{|\mathcal{K}(g)|}} & & \downarrow{\scriptstyle \overline{g'}} \\
|\mathcal{K}(Y)|/|\mathcal{K}(B)| & \xrightarrow{hf_{\mathcal{K}(Y)/\mathcal{K}(B)}} & Y'/B'
\end{array}
$$

For the first square this is clear, since the convenient cell structures for $|\mathcal{K}(X)|$ and $|\mathcal{K}(Y)|$ induce convenient cell structures for the subcomplexes $|\mathcal{K}(A)|$ and $|\mathcal{K}(B)|$ and in this case $f_{\mathcal{K}(A)} = \mu_{\mathcal{K}(A)}$ and $f_{\mathcal{K}(B)} = \mu_{\mathcal{K}(B)}$ by Proposition 7.1.11. For the second square we just have to remember that there exists a homotopy $H : \mu_{\mathcal{K}(Y)}|\mathcal{K}(g)| \simeq g'\mu_{\mathcal{K}(X)}$ such that $H(|\mathcal{K}(A)| \times I) \subseteq B'$ by Remark 1.4.14 and this induces a homotopy $\overline{H} : |\mathcal{K}(X)|/|\mathcal{K}(A)| \times I \to Y'/B'$ which is the homotopy between $hf_{\mathcal{K}(Y)/\mathcal{K}(B)}\overline{|\mathcal{K}(g)|}$ and $\overline{g'}hf_{\mathcal{K}(X)/\mathcal{K}(A)}$ by Proposition 7.1.11 and Proposition 7.2.4. $\square$

Remark 7.2.6. There is an alternative, and perhaps simpler, way to prove the existence of the sequence (7.1) and Corollary 7.2.5 which does not use the fact that $\mathcal{X}(K)/\mathcal{X}(L)$ is a finite model of $|K|/|L|$ when L is a full subcomplex of K. Nevertheless we chose the proof of above because it provides an explicit formula for the weak homotopy equivalence $|K|/|L| \to \mathcal{X}(K)/\mathcal{X}(L)$.

The idea of the alternative proof is as follows: if L is a full subcomplex of K, $\mathcal{X}(L)^{op}$ is a closed subspace of $\mathcal{X}(K)^{op}$ which is a deformation retract of its open hull $\mathcal{X}(L)^{op} \subseteq \mathcal{X}(K)^{op}$. Therefore, there is a long exact sequence as in Proposition 7.1 but for the opposite spaces $\mathcal{X}(L)^{op}, \mathcal{X}(K)^{op}$ and $\mathcal{X}(K)^{op}/\mathcal{X}(L)^{op}$. Using the associated complexes of these spaces we obtain the long exact sequence of Proposition 7.1 and the naturality of Corollary 7.2.5 follows from the naturality of the sequence for the opposite spaces.

Chapter 8
Group Actions and a Conjecture of Quillen

In his seminal article [70], Daniel Quillen studied algebraic properties of a
finite group by means of homotopy properties of a certain complex $\mathcal{K}(S_p(G))$
associated to the group. Given a finite group G and a prime integer p dividing
the order of G, let $S_p(G)$ denote the poset of nontrivial p-subgroups of G
ordered by inclusion. The poset $S_p(G)$, or more concretely, its associated
simplicial complex $\mathcal{K}(S_p(G))$, was first investigated by Brown. In his 1975
paper [20], Brown proved a very interesting variation of Sylow's Theorems
for the Euler characteristic. In [70] Quillen delved deeper into the topological
properties of $\mathcal{K}(S_p(G))$ and their relationship with the algebraic properties
of G. He showed, for instance, that if G has a nontrivial normal p-subgroup,
$|\mathcal{K}(S_p(G))|$ is contractible. He proved that the converse of this statement is
true for solvable groups and conjectured that it is true for all finite groups.
Important advances were made in [3] but a complete answer to Quillen's
question is still unknown.

Apparently, Brown and Quillen were not aware of the theory of finite
spaces at that time. They worked with the associated complex $\mathcal{K}(S_p(G))$
without considering the intrinsic topology of the poset $S_p(G)$. Stong was the
first mathematician who studied Quillen's conjecture from the viewpoint of
finite spaces. In [77], he developed the equivariant homotopy theory of finite
spaces and studied its relationship with the conjecture. Stong showed that
the group G has a nontrivial p-subgroup if and only if $S_p(G)$ is a contractible
finite space. In view of this result, the conjecture can be restated as:

$S_p(G)$ is contractible if and only if it is homotopically trivial.

In this chapter we will recall Stong's equivariant homotopy theory for finite
spaces and its connection with Quillen's conjecture. Then we will develop an
equivariant simple homotopy theory for complexes and finite spaces that
allows one to study the conjecture from a new point of view.

J.A. Barmak, *Algebraic Topology of Finite Topological Spaces and*
Applications, Lecture Notes in Mathematics 2032,
DOI 10.1007/978-3-642-22003-6_8, © Springer-Verlag Berlin Heidelberg 2011

8.1　Equivariant Homotopy Theory for Finite Spaces

Let G be a group. By a G-*space* we will mean a topological space X with an action of G such that the maps $m_g : X \to X$ defined by $m_g(x) = gx$ are continuous for every $g \in G$. A G-*map* (or *equivariant map*) between G-spaces X and Y is a continuous map $f : X \to Y$ such that $f(gx) = gf(x)$ for every $g \in G$ and $x \in X$. A homotopy $H : X \times I \to Y$ is a G-*homotopy* (or *equivariant homotopy*) if $H(gx, t) = gH(x, t)$ for every $g \in G, x \in X, t \in I$. A G-map $f : X \to Y$ is a G-*homotopy equivalence* if there exists a G-map $h : Y \to X$ and G-homotopies between hf and 1_X and between fh and 1_Y. A subspace A of a G-space X is said to be G-*invariant* if $ga \in A$ for every $g \in G, a \in A$. A G-invariant subspace $A \subseteq X$ is an *equivariant strong deformation retract* if there is an equivariant retraction $r : X \to A$ such that ir is homotopic to 1_X via a G-homotopy which is stationary at A.

If x is a point of a G-space X, $Gx = \{gx\}_{g \in G}$ denotes the orbit of x. The set of points fixed by the action is denoted by $X^G = \{x \in X \mid gx = x \, \forall g \in G\}$.

A finite T_0-space which is a G-space will be called a finite T_0-G-space.

The following is a general result about automorphisms of posets but we will need it only in the context of finite T_0-G-spaces.

Lemma 8.1.1. *Let X be a finite T_0-space, $x \in X$ and let $f : X \to X$ be an automorphism. If x and $f(x)$ are comparable, $x = f(x)$. Moreover, if $f_1, f_2 : X \to X$ are two automorphisms and $f_1(x)$ is comparable with $f_2(x)$, then $f_1(x) = f_2(x)$.*

Proof. Assume without loss of generality that $x \leq f(x)$. Then $f^i(x) \leq f^{i+1}(x)$ for every $i \geq 0$. By the finiteness of X, the equality must hold for some i and since f is a homeomorphism $x = f(x)$. The second part of the lemma follows from the first by considering the automorphism $f_2^{-1} f_1$. □

Lemma 8.1.2. *Let X be a finite T_0-G-space. Then there exists a core of X which is G-invariant and an equivariant strong deformation retract of X.*

Proof. Suppose X is not minimal. Then there exists a beat point $x \in X$. Without loss of generality suppose x is a down beat point that covers a point y. The orbit of x and the orbit of y are disjoint. If $gx = hy$, then $gx = hy < hx$, which contradicts Lemma 8.1.1. Moreover, if $gx = hx$, then $h^{-1}gy < h^{-1}gx = x$ and therefore $h^{-1}gy \leq y$. By Lemma 8.1.1, $h^{-1}gy = y$ and then $gy = hy$. Therefore, the retraction $r : X \to X \smallsetminus Gx$ defined by $r(gx) = gy$ is a well defined continuous G-map. The homotopy $X \times I \to X$ corresponding to the path $\alpha : I \to X^X$ given by $\alpha(t) = ir$ if $0 \leq t < 1$ and $\alpha(1) = 1_X$ is a G-homotopy between ir and 1_X relative to $X \smallsetminus Gx$. Therefore $X \smallsetminus Gx$ is an equivariant strong deformation retract of X. The proof is concluded by an inductive argument. □

Proposition 8.1.3. *A contractible finite T_0-G-space has a point which is fixed by the action of G.*

Proof. By Lemma 8.1.2 there is a core, i.e. a point, which is G-invariant.   □

Example 8.1.4. Let G be a finite group and suppose there exists a proper subgroup $H \subsetneq G$ such that for every nontrivial subgroup S of G, $S \cap H$ is nontrivial. Then G is not a simple group.

   Although at first sight this result does not seem to be related to finite spaces, we will give a proof using Proposition 8.1.3. Since H is a proper subgroup of G, G is nontrivial and therefore $H = G \cap H$ is nontrivial. Consider the poset $S(G)$ of nontrivial proper subgroups of G. Let $c_H : S(G) \to S(G)$ be the constant map H and define $f : S(G) \to S(G)$ by $f(S) = S \cap H$. The map f is well defined by hypothesis and it is clearly continuous. Moreover, $1_{S(G)} \geq f \leq c_H$ and then $S(G)$ is contractible. On the other hand, G acts on $S(G)$ by conjugation. Then, by Proposition 8.1.3, G has a nontrivial proper normal subgroup.

   For instance, let $\mathcal{Q} = \{1, -1, i, -i, j, -j, k, -k\}$ be the quaternion group, where $(-1)^2 = 1, (-1)i = i(-1) = -i, (-1)j = j(-1) = -j, (-1)k = k(-1) = -k, i^2 = j^2 = k^2 = ijk = -1$. Let $H = \{1, -1\}$. Then H is in the hypothesis of the statement since -1 is a power of every nontrivial element of $\mathcal{Q}$. Therefore, $\mathcal{Q}$ is not simple.

   There are also purely algebraic (and simple) proofs of this result. In fact is easy to see that in the hypothesis of above, $\bigcap_{g \in G} gHg^{-1}$ is a nontrivial normal subgroup of G.

   Proposition 8.1.3 cannot be generalized to non-finite spaces. The analogous statement for simplicial complexes is not true. If K is a contractible finite simplicial complex with a simplicial action of a finite group G, then it may be the case that there is no point fixed by the induced action in $|K|$. Moreover, Oliver [64] gave a description of the groups G for which there exists a simplicial fixed point free action on a contractible simplicial complex. However we can prove that every simplicial action on a strong collapsible complex has a fixed point. The following result appears in [11].

Theorem 8.1.5. *Let K be a strong collapsible simplicial complex and let G be a group acting simplicially on K. Then there is a point in $|K|$ which is fixed by the action induced in the geometric realization.*

Proof. The action on K induces and action on $\mathcal{X}(K)$, which is contractible by Theorem 5.2.2. By Proposition 8.1.3, there is a point of $\mathcal{X}(K)$ fixed by the action. This is a G-invariant simplex of K, and therefore its barycenter is a fixed point of the corresponding action on $|K|$.   □

Proposition 8.1.6. *Let X and Y be finite T_0-G-spaces and let $f : X \to Y$ be a G-map which is a homotopy equivalence. Then f is an equivariant homotopy equivalence.*

Proof. Let X_c and Y_c be cores of X and Y which are equivariant strong deformation retracts. Denote i_X, i_Y and r_X, r_Y the inclusions and equivariant strong deformation retractions. Since f is a homotopy equivalence and a G-map, so is $r_Y f i_X : X_c \to Y_c$. Therefore, $r_Y f i_X$ is a G-isomorphism. Define the G-map $g = i_X (r_Y f i_X)^{-1} r_Y : Y \to X$, then

$$ fg = f i_X (r_Y f i_X)^{-1} r_Y \simeq i_Y r_Y f i_X (r_Y f i_X)^{-1} r_Y = i_Y r_Y \simeq 1_Y, $$

$$ gf = i_X (r_Y f i_X)^{-1} r_Y f \simeq i_X (r_Y f i_X)^{-1} r_Y f i_X r_X = i_X r_X \simeq 1_X. $$

All the homotopies being equivariant. Therefore f is an equivariant homotopy equivalence with homotopy inverse g.   □

Remark 8.1.7. Two finite T_0-G-spaces which are homotopy equivalent, need not have the same equivariant homotopy type. Let $X = \mathbb{S}(S^0)$. The group of automorphisms $Aut(X)$ acts on X in the usual way by $f \cdot x = f(x)$ and in the trivial way by $f \circ x = x$. Denote by X_0 the $Aut(X)$-space with the first action and by X_1, the second. Suppose there exists an equivariant homotopy equivalence $g : X_0 \to X_1$. Since X is minimal, g is a homeomorphism. Let $f : X \to X$ be an automorphism distinct from the identity. Then $gf(x) = g(f \cdot x) = f \circ g(x) = g(x)$ for every $x \in X$. Thus, $f = 1_X$, which is a contradiction.

8.2   The Poset of Nontrivial p-Subgroups and the Conjecture of Quillen

In this section, we recall Quillen's basic results on the poset $S_p(G)$ and the poset $A_p(G)$ of elementary abelian p-subgroups. We will recall also Stong's reformulation of the conjecture for finite spaces and we will exhibit an alternative proof of K. Brown's result on the Euler characteristic of $S_p(G)$.

In the following, G will denote a finite group and p a prime integer dividing the order of G. The elements of the poset $S_p(G)$ are the nontrivial p-subgroups of G, namely the subgroups different from the trivial subgroup of one element, whose order is a power of p. Note that the maximal elements of $S_p(G)$ are the Sylow p-subgroups of G and the minimal elements correspond to the subgroups of order p.

Example 8.2.1. For $G = D_6 = \langle s, r \mid s^2 = r^6 = srsr = 1 \rangle$, the dihedral group of order 12, and $p = 2$, the poset $S_2(D_6)$ looks as follows

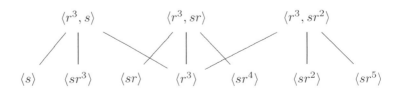

Quillen proved in [70] that if G has a nontrivial normal p-subgroup then the polyhedron $|\mathcal{K}(S_p(G))|$ is contractible. But with his proof it can be deduced that $S_p(G)$ is a contractible finite space, which a priori is stronger.

Theorem 8.2.2 (Quillen). *If G has a nontrivial normal p-subgroup, $S_p(G)$ is contractible.*

Proof. Suppose N is a nontrivial normal p-subgroup of G. Define $f : S_p(G) \to S_p(G)$ by $f(H) = NH = \{nh \mid n \in N, \ h \in H\}$. NH is a subgroup of G since $N \lhd G$. Moreover, NH is a quotient of the semidirect product $N \rtimes H$, where $(n_1, h_1)(n_2, h_2) = (n_1 h_1 n_2 h_1^{-1}, h_1 h_2)$. Since N and H are p-groups, so is NH. Therefore, f is well defined. Clearly f is order preserving, and if c_N denotes the constant map N, $c_N \leq f \geq 1_{S_p(G)}$. Thus $1_{S_p(G)}$ is homotopic to a constant and then, $S_p(G)$ is contractible. $\qquad\square$

Corollary 8.2.3. *If G has a nontrivial normal p-subgroup, $|\mathcal{K}(S_p(G))|$ is contractible.*

The conjecture of Quillen is the converse of this result.

Conjecture 8.2.4 (Quillen). *If $|\mathcal{K}(S_p(G))|$ is contractible, G has a nontrivial normal p-subgroup.*

Following Stong [77], we will use the results of the previous section to give a restatement of the conjecture in the setting of finite spaces.

Theorem 8.2.5 (Stong). *Let G be a finite group and let p be a prime integer. Then $S_p(G)$ is contractible if and only if G has a nontrivial normal p-subgroup.*

Proof. In view of Theorem 8.2.2 we only need to prove the existence of a nontrivial normal p-subgroup provided that $S_p(G)$ is contractible. The poset $S_p(G)$ is a G-space with the action given by conjugation, $g \cdot H = gHg^{-1}$. If $S_p(G)$ is contractible, by Proposition 8.1.3, there exists $N \in S_p(G)$ such that $gNg^{-1} = N$ for every $g \in G$, i.e. N is a normal subgroup of G. $\qquad\square$

In light of Theorem 8.2.5, the conjecture may be restated as follows:

Restatement of Quillen's conjecture (Stong): If $S_p(G)$ is homotopically trivial, it is contractible.

In [70], Quillen shows that his conjecture 8.2.4 is true for solvable groups. To do this, Quillen works with another poset $A_p(G)$ which is weak homotopy equivalent to $S_p(G)$, and proves that if G does not have nontrivial normal p-subgroups, then $A_p(G)$ has a nonvanishing homology group. The finite space $A_p(G)$ is the subspace of $S_p(G)$ consisting of the elementary abelian p-subgroups, i.e. abelian subgroups whose elements have all order 1 or p.

Proposition 8.2.6. *The inclusion $A_p(G) \hookrightarrow S_p(G)$ is a weak homotopy equivalence.*

Proof. By Theorem 1.4.2, it suffices to show that $i^{-1}(U_H) = A_p(H)$ is contractible for every $H \in S_p(G)$. Since H is a nontrivial p-subgroup, its center Z is not trivial. Let $N \subseteq Z$ be the subgroup of elements of order 1 or p. If $T \in A_p(H)$, $TN \in A_p(H)$ and $T \leq TN \geq N$. Therefore, $A_p(H)$ is contractible. $\qquad\square$

In [77], Stong shows that in general $A_p(G)$ and $S_p(G)$ are not homotopy equivalent, however, if $A_p(G)$ is contractible, there is a point fixed by the action of G and then $S_p(G)$ is contractible. Apparently it is unknown whether the converse of this result holds.

Example 8.2.7. Let Σ_5 be the symmetric group on five letters. We give an alternative proof of the well known fact that Σ_5 has no nontrivial normal 2-subgroups.

The subgroup $\langle (1234), (13) \rangle \subseteq \Sigma_5$ has order 8 and it is not abelian. All the other subgroups of order 8 are isomorphic to this Sylow 2-subgroup and therefore, Σ_5 has no elementary abelian subgroups of order 8. Thus, the height of the poset $A_2(\Sigma_5)$ is at most one.

On the other hand, there is a subspace of $A_2(\Sigma_5)$ with the following Hasse diagram

Then the graph $\mathcal{K}(A_2(\Sigma_5))$ has a cycle and therefore it is not contractible. Hence, $A_2(\Sigma_5)$ is not homotopically trivial and then neither is $S_2(\Sigma_5)$. In particular, $S_2(\Sigma_5)$ is not contractible and then Σ_5 does not have normal 2-subgroups which are nontrivial.

Now we exhibit an alternative proof of K. Brown's result on Euler characteristic [20]. If H is a subgroup of G, then it also acts on $S_p(G)$ by conjugation and $S_p(G)^H$ denotes then the fixed point set of this action.

Proposition 8.2.8. *Let H be a nontrivial p-subgroup of G. Then $S_p(G)^H$ is contractible.*

Proof. If $T \in S_p(G)^H$, $TH \in S_p(G)^H$. Since $T \leq TH \geq H$, the constant map $c_H : S_p(G)^H \to S_p(G)^H$ is homotopic to the identity. $\qquad\square$

Note that if X is a finite T_0-G-space, the subdivision X' is also a G-space with the action given by $g \cdot \{x_0, x_1, \ldots, x_n\} = \{gx_0, gx_1, \ldots, gx_n\}$.

Let P be a Sylow p-subgroup of G. The action of P on $S_p(G)$ by conjugation induces an action of P on $S_p(G)'$. Given $c \in S_p(G)'$, let $P_c =$

$\{g \in P \mid gc = c\}$ denote the isotropy group (or stabilizer) of c. Define $Y = \{c \in S_p(G)' \mid P_c \neq 0\}$.

Lemma 8.2.9. $\chi(S_p(G)', Y) \equiv 0 \ mod(\#P)$.

Proof. Let $C = \{c_0 < c_1 < \ldots < c_n\} \in S_p(G)''$ be a chain of $S_p(G)'$ which is not a chain of Y. Then there exists $0 \leq i \leq n$ such that $c_i \notin Y$. Therefore, if g and h are two different elements of P, $gc_i \neq hc_i$. In other words, the orbit of c_i under the action of P has $\#P$ elements. Thus, the orbit of C also has $\#P$ elements. In particular, $\#P$ divides $\chi(S_p(G)', Y) = \sum_{i \geq 0}(-1)^i \alpha_i$, where α_i is the number of chains of $(i+1)$-elements of $S_p(G)'$ which are not chains of Y. $\qquad \square$

Lemma 8.2.10. Y *is homotopically trivial.*

Proof. Let $f : Y \to S_p(P)^{op}$ defined by $f(c) = P_c$, the isotropy group of c. By definition of Y, P_c is a nontrivial subgroup of P and then f is a well defined function. If $c_0 \leq c_1$, $P_{c_0} \supseteq P_{c_1}$. Thus, f is continuous. If $0 \neq H \subseteq P$, $f^{-1}(U_H) = \{c \in Y \mid H \subseteq P_c\} = (S_p(G)^H)'$, which is contractible by Proposition 8.2.8. From Theorem 1.4.2 we deduce that f is a weak homotopy equivalence. Since $S_p(P)^{op}$ has minimum, Y is homotopically trivial. $\qquad \square$

In [70], Quillen proves that Y is homotopically trivial finding a third space Z which is weak homotopy equivalent to Y and $S_p(P)$. Our proof is somewhat more direct.

Theorem 8.2.11 (Brown). $\chi(S_p(G)) \equiv 1 \ mod(\#P)$.

Proof. Since $\chi(Y) = 1$ by Lemma 8.2.10, $\chi(S_p(G)) = \chi(S_p(G)') = \chi(Y) + \chi(S_p(G)', Y) \equiv 1 \ mod(\#P)$. $\qquad \square$

8.3 Equivariant Simple Homotopy Types

To prove Quillen's conjecture, one would need to show that if $S_p(G)$ is homotopically trivial, then the action of G by conjugation has a fixed point. However there exist homotopically trivial finite T_0-G-spaces without fixed points. To construct such an example it is enough to take a contractible finite simplicial complex with a fixed point free action [64] and consider the associated finite space.

The proof of Proposition 8.1.3 and the previous results suggest that the hypothesis of contractibility can be replaced by a weaker notion. Combining these ideas with the simple homotopy theory of finite spaces, we introduce the notion of G-collapse of finite spaces and of simplicial complexes. These two concepts are strongly related, similarly to the non-equivariant case.

Equivariant simple homotopy types of finite spaces allow us to attack the conjecture of Quillen and to deepen our understanding of equivariant homotopy theory of finite spaces originally studied by Stong. In this section we will only develop the simple and strong equivariant homotopy theory for finite spaces and complexes. Applications to the poset of p-subgroups appear in the next section.

As in the previous section, G will denote a finite group.

Recall that there is a strong collapse from a finite T_0-space X to a subspace Y if the second one is obtained from the first by removing beat points. By our results on minimal pairs (Sect. 2.2), this is equivalent to saying that $Y \subseteq X$ is a strong deformation retract. We denote this situation by $X \searrow\!\!\!\!\searrow Y$.

If x is a beat point of a finite T_0-G-space X, $gx \in X$ is a beat point for every $g \in G$. In this case we say that there is an *elementary strong G-collapse* from X to $X \smallsetminus Gx$. Note that elementary strong G-collapses are strong collapses. A sequence of elementary strong G-collapses is called a *strong G-collapse* and it is denoted by $X \searrow\!\!\!\!\searrow^G Y$. Strong G-expansions are dually defined.

Proposition 8.3.1. *Let X be a finite T_0-G-space and $Y \subseteq X$ a G-invariant subspace. The following are equivalent:*

  i. $X \searrow\!\!\!\!\searrow^G Y$.
 ii. $Y \subseteq X$ *is an equivariant strong deformation retract.*
iii. $Y \subseteq X$ *is a strong deformation retract.*

Proof. If there is an elementary strong G-collapse from X to Y, then by the proof of Lemma 8.1.2, Y is an equivariant strong deformation retract of X.

If $Y \subseteq X$ is a strong deformation retract and $x \in X \smallsetminus Y$ is a beat point of X, $X \searrow\!\!\!\!\searrow^G X \smallsetminus Gx = X_1$. In particular $X_1 \subseteq X$ is a strong deformation retract, and then, so is $Y \subseteq X_1$. By induction, $X_1 \searrow\!\!\!\!\searrow^G Y$ and then $X \searrow\!\!\!\!\searrow^G Y$.
□

Let X be a finite T_0-G-space. A core of X which is G-invariant is called a *G-core*. From Stong's results (Lemma 8.1.2), it follows that every finite T_0-G-space has a G-core.

Definition 8.3.2. Let X be a finite T_0-G-space. If $x \in X$ is a weak point, $gx \in X$ is also a weak point for every $g \in G$ and we say that there is an *elementary G-collapse* from X to $X \smallsetminus Gx$. This is denoted by $X \searrow^{Ge} X \smallsetminus Gx$. Note that the resulting subspace $X \smallsetminus Gx$ is G-invariant. A sequence of elementary G-collapses is called a *G-collapse* and it is denoted $X \searrow^G Y$. *G-expansions* are defined dually. X is *G-collapsible* if it G-collapses to a point.

Note that strong G-collapses are G-collapses and that G-collapses are collapses. If the action is trivial, G-collapses and collapses coincide.

A finite T_0-G-space is strong collapsible if and only if it is G-strong collapsible. However, this is not true for collapsibility and G-collapsibility as the next example shows.

Example 8.3.3. Let X be the following finite space (cf. Fig. 7.3 in page 97)

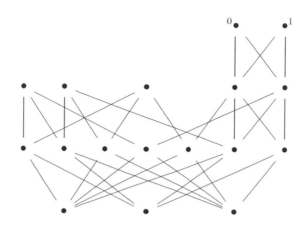

Consider the action of the two-element group $\mathbb{Z}_2$ over X that permutes 0 and 1 and fixes every other element. The unique weak points of X are 0 and 1. $X \smallsetminus \{0\}$ is collapsible but $X \smallsetminus \{0, 1\}$ is not. Therefore X is a collapsible finite space which is not G-collapsible.

The notion of G-collapse can be studied also in the setting of simplicial complexes. Suppose K is a finite G-simplicial complex and $\sigma \in K$ is a free face of $\sigma' \in K$. Then for every $g \in G$, $g\sigma$ is a free face of $g\sigma'$, however it is not necessarily true that K collapses to $K \smallsetminus \bigcup_{g \in G} \{g\sigma, g\sigma'\}$.

Example 8.3.4. Let σ' be a 2-simplex and $\sigma \subsetneqq \sigma'$ a 1-face of σ. Consider the action of $\mathbb{Z}_3$ by rotation over $K = \overline{\sigma'}$. Then σ is a free face of σ', but $\overline{\sigma'}$ does not collapse to $\overline{\sigma'} \smallsetminus \bigcup_{g \in \mathbb{Z}_3} \{g\sigma, g\sigma'\}$ which is the discrete complex with 3 vertices.

If σ is a free face of σ' in the G-complex K, and $g \in G$ is such that $g\sigma = \sigma$, then $\sigma \subsetneqq g\sigma'$ and therefore $g\sigma' = \sigma'$. In other words, the isotropy group G_σ of σ is contained in the isotropy group $G_{\sigma'}$ of σ'. The other inclusion does not hold in general as the previous example shows.

Definition 8.3.5. Let K be a finite G-simplicial complex and let $\sigma \in K$ be a free face of $\sigma' \in K$ ($\sigma \subsetneqq \sigma'$ is a *collapsible pair*). Consider the G-invariant subcomplex $L = K \smallsetminus \bigcup_{g \in G} \{g\sigma, g\sigma'\}$. We say that there is an *elementary G-collapse* $K \searrow^{Ge} L$ from K to L, or that $\sigma \subsetneqq \sigma'$ is a *G-collapsible pair*, if $G_\sigma = G_{\sigma'}$. A sequence of elementary G-collapses is called a *G-collapse* and denoted by $K \searrow^{G} L$. A G-complex K is *G-collapsible* if it G-collapses to a vertex.

Proposition 8.3.6. *Let K be a finite G-simplicial complex and let $\sigma \subsetneq \sigma'$ be a collapsible pair. The following are equivalent:*

1. $\sigma \subsetneq \sigma'$ *is a G-collapsible pair.*
2. $K \searrow L = K \smallsetminus \bigcup_{g \in G} \{g\sigma, g\sigma'\}.$

Proof. Suppose σ is an n-simplex and that $K \searrow L$. Then the set $\bigcup_{g \in G} \{g\sigma, g\sigma'\}$ contains as many n-simplices as $(n+1)$-simplices, i.e. the sets $G \cdot \sigma = \{g\sigma\}_{g \in G}$ and $G \cdot \sigma' = \{g\sigma'\}_{g \in G}$ have the same cardinality. Therefore

$$\#G_\sigma = \#G/\#G \cdot \sigma = \#G/\#G \cdot \sigma' = \#G_{\sigma'}.$$

Since $G_\sigma \subseteq G_{\sigma'}$, the equality holds.

Conversely, suppose $\sigma \subsetneq \sigma'$ is a G-collapsible pair. Then the pairs $g\sigma \subsetneq g\sigma'$ can be collapsed one by one.   □

Therefore, G-collapses are collapses. The following is an extension of the classical result of Whitehead, Proposition 4.1.3, which says that if $K_1, K_2 \subseteq K$ are finite simplicial complexes, then $K_1 \cup K_2 \searrow K_1$ if and only if $K_2 \searrow K_1 \cap K_2$ (with the same sequence of collapses). The proof is straightforward.

Remark 8.3.7. Let K be a finite G-simplicial complex and let $K_1, K_2 \subseteq K$ be two G-invariant subcomplexes such that $K_1 \cup K_2 = K$. Then, $K \searrow^G K_1$ if and only if $K_2 \searrow^G K_1 \cap K_2$.

Remark 8.3.8. Let X be a finite T_0-G-space. If X is G-collapsible, it collapses to a G-invariant one-point subspace. In particular, the fixed point set X^G is nonempty.

Now we will study the relationship between G-collapses of finite spaces and simplicial G-collapses.

If X is a finite T_0-G-space, there is a natural induced action on $\mathcal{K}(X)$. If we consider G both as a discrete topological group and a discrete simplicial complex, there is a natural isomorphism $\mathcal{K}(G \times X) = G \times \mathcal{K}(X)$ and an action $\theta : G \times X \to X$ induces an action $\mathcal{K}(\theta) : G \times \mathcal{K}(X) = \mathcal{K}(G \times X) \to \mathcal{K}(X)$ Analogously, an action $\theta : G \times K \to K$ over a finite simplicial complex K induces an action $\mathcal{X}(\theta) : G \times \mathcal{X}(K) = \mathcal{X}(G \times K) \to \mathcal{X}(K)$.

Unless we say the opposite, if X is a finite T_0-G-space and K a finite G-simplicial complex, we will assume the actions over $\mathcal{K}(X)$ and $\mathcal{X}(K)$ are the induced ones.

The main aim of this section is to prove the equivariant version of Theorem 4.2.11. The proof will be similar to the proof of the non-equivariant case.

Lemma 8.3.9. *Let aK be a finite simplicial cone and suppose G acts on aK fixing the vertex a. Then $aK \searrow^G a$.*

Proof. Let σ be a maximal simplex of K. Then $\sigma \subsetneq a\sigma$ is a G-collapsible pair since $g \cdot a\sigma = a\sigma$ implies $g\sigma = \sigma$. Therefore $aK \searrow^G aK \smallsetminus \bigcup_{g \in G} \{g\sigma, g \cdot a\sigma\} = a(K \smallsetminus \bigcup_{g \in G} \{g\sigma\})$. The lemma follows from an inductive argument. $\square$

Lemma 8.3.10. *Let X be a finite T_0-G-space and let $x \in X$. The stabilizer G_x of x acts on $\hat{C}_x$ and then on $\mathcal{K}(\hat{C}_x)$. If $\mathcal{K}(\hat{C}_x)$ is G_x-collapsible, $\mathcal{K}(X) \searrow^G \mathcal{K}(X \smallsetminus Gx)$.*

Proof. If $\sigma \subsetneq \sigma'$ is a G_x-collapsible pair in $\mathcal{K}(\hat{C}_x)$, $x\sigma \subsetneq x\sigma'$ is G_x-collapsible in $x\mathcal{K}(\hat{C}_x)$. In this way, copying the elementary G_x-collapses of $\mathcal{K}(\hat{C}_x) \searrow^{G_x} *$, one obtains that $\mathcal{K}(C_x) = x\mathcal{K}(\hat{C}_x) \searrow^{G_x} \mathcal{K}(\hat{C}_x) \cup \{x, x*\} \searrow^{G_x} \mathcal{K}(\hat{C}_x)$. Now we will show that since $\mathcal{K}(C_x) \searrow^{G_x} \mathcal{K}(\hat{C}_x)$,

$$\bigcup_{g \in G} g\mathcal{K}(C_x) \searrow^G \bigcup_{g \in G} g\mathcal{K}(\hat{C}_x). \tag{8.1}$$

Suppose $\mathcal{K}(C_x) = K_0 \searrow^{G_x e} K_1 \searrow^{G_x e} K_2 \searrow^{G_x e} \ldots \searrow^{G_x e} K_r = \mathcal{K}(\hat{C}_x)$. Notice that all the simplices removed in these collapses contain the vertex x. If $\sigma \subsetneq \sigma'$ is the G_x-collapsible pair collapsed in $K_i \searrow^{G_x e} K_{i+1}$ (along with the other simplices in the orbits of σ and σ'), we claim that $\sigma \subsetneq \sigma'$ is G-collapsible in $\bigcup_{g \in G} gK_i$. Suppose $\sigma \subsetneq g\tilde{\sigma}$ with $g \in G$, $\tilde{\sigma} \in K_i$. Since $x \in \sigma \subsetneq g\tilde{\sigma}$, $g^{-1}x \in \tilde{\sigma}$ and then x and $g^{-1}x$ are comparable. By Lemma 8.1.1 $x = g^{-1}x$ and therefore $g \in G_x$. Since K_i is G_x-invariant and σ is a free face of σ' in K_i, $g\tilde{\sigma} = \sigma'$. Therefore, $\sigma \subsetneq \sigma'$ is a collapsible pair in $\bigcup_{g \in G} gK_i$.

Let $g \in G$ be such that $g\sigma' = \sigma'$. By the same argument as above, x, $gx \in \sigma'$ and then $g \in G_x$. Since $\sigma \subsetneq \sigma'$ is G_x-collapsible in K_i, $g\sigma = \sigma$, which proves that it is also G-collapsible in $\bigcup_{g \in G} gK_i$. Thus,

$$\bigcup_{g \in G} gK_i \searrow^{Ge} \bigcup_{g \in G} gK_i \smallsetminus \bigcup_{g \in G} \{g\sigma, g\sigma'\} = \bigcup_{g \in G} (gK_i \smallsetminus \bigcup_{h \in G} \{gh\sigma, gh\sigma'\})$$

$$= \bigcup_{g \in G} g(K_i \smallsetminus \bigcup_{h \in G} \{h\sigma, h\sigma'\}).$$

But $h\sigma$ and $h\sigma'$ are simplices of K_i if and only if $h \in G_x$, then

$$\bigcup_{g \in G} g(K_i \smallsetminus \bigcup_{h \in G} \{h\sigma, h\sigma'\}) = \bigcup_{g \in G} g(K_i \smallsetminus \bigcup_{h \in G_x} \{h\sigma, h\sigma'\}) = \bigcup_{g \in G} gK_{i+1}.$$

So (8.1) is proved, i.e.

$$\bigcup_{g\in G} g\mathcal{K}(C_x) \searrow^G \bigcup_{g\in G} g\mathcal{K}(\hat{C}_x) = (\bigcup_{g\in G} g\mathcal{K}(C_x)) \cap \mathcal{K}(X \smallsetminus Gx).$$

By Remark 8.3.7,

$$\mathcal{K}(X) = (\bigcup_{g\in G} g\mathcal{K}(\hat{C}_x)) \cup \mathcal{K}(X \smallsetminus Gx) \searrow^G \mathcal{K}(X \smallsetminus Gx).$$

$\square$

Theorem 8.3.11.

(a) *Let X be a finite T_0-G-space and $Y \subseteq X$ a G-invariant subspace. If $X \searrow^G Y$, $\mathcal{K}(X) \searrow^G \mathcal{K}(Y)$.*

(b) *Let K be a finite G-simplicial complex and $L \subseteq K$ a G-invariant subcomplex. If $K \searrow^G L$, $\mathcal{X}(K) \searrow^G \mathcal{X}(K)$.*

Proof. Suppose first that $x \in X$ is a beat point. Then there exists $y \in X$, $y \neq x$ such that $C_x \subseteq C_y$. Therefore $G_x \subseteq G_y$ by Lemma 8.1.1 and $\mathcal{K}(\hat{C}_x) = y\mathcal{K}(\hat{C}_x \smallsetminus \{y\})$. The stabilizer G_x of x acts on $\hat{C}_x$, and therefore on $\mathcal{K}(\hat{C}_x)$, and fixes y. By Lemma 8.3.9, $\mathcal{K}(\hat{C}_x) \searrow^{G_x} y$. By Lemma 8.3.10, $\mathcal{K}(X) \searrow^G \mathcal{K}(X \smallsetminus Gx)$. In particular if X is contractible, this says that $\mathcal{K}(X)$ is G-collapsible.

Suppose now that $x \in X$ is a weak point. Then C_x is contractible and $\mathcal{K}(C_x)$ is G_x-collapsible. Again from Lemma 8.3.10, we obtain that $\mathcal{K}(X) \searrow^G \mathcal{K}(X \smallsetminus Gx)$. This proves the first part of the theorem for elementary G-collapses. The general case follows immediately from this one.

To prove the second part of the theorem we can suppose that K elementary G-collapses to L. Let $\sigma \subsetneq \sigma'$ be a G-collapsible pair in K such that $L = K \smallsetminus \{g\sigma, g\sigma'\}_{g\in G}$. Then, $\sigma \in \mathcal{X}(K)$ is an up beat point and therefore $\mathcal{X}(K) \searrow^{Ge} \mathcal{X}(K) \smallsetminus \{g\sigma\}_{g\in G}$. Now, $\sigma' \in \mathcal{X}(K) \smallsetminus \{gS\}_{g\in G}$ is a down weak point since $\overline{\sigma'} \smallsetminus \{\sigma, \sigma'\}$ is a simplicial cone and then $\hat{U}_{\sigma'}^{\mathcal{X}(K) \smallsetminus \{g\sigma\}_{g\in G}} = \hat{U}_{\sigma'}^{\mathcal{X}(K) \smallsetminus \{\sigma\}} = \mathcal{X}(\overline{\sigma'} \smallsetminus \{\sigma, \sigma'\})$ is contractible by Lemma 4.2.6. Therefore, $\mathcal{X}(K) \smallsetminus \{g\sigma\}_{g\in G} \searrow^{Ge} \mathcal{X}(K) \smallsetminus \{g\sigma, g\sigma'\}_{g\in G} = \mathcal{X}(L)$ and $\mathcal{X}(K) \searrow^G \mathcal{X}(L)$.

$\square$

The equivalence classes of the equivalence relations $\nearrow\!\!\!\diagdown^G$ generated by the G-collapses are called *equivariant simple homotopy types* in the setting of finite spaces and of simplicial complexes. An easy modification of Proposition 4.2.9 shows that if X is a finite T_0-G-space, X and X' are equivariantly simple homotopy equivalent (see Proposition 8.3.21). Therefore, we have the following Corollary of Theorem 8.3.11.

Corollary 8.3.12. *Let X and Y be finite T_0-G-spaces. Then X and Y have the same equivariant simple homotopy type if and only if $\mathcal{K}(X)$ and $\mathcal{K}(Y)$ have the same equivariant simple homotopy type.*

However, the analogous result for the functor $\mathcal{X}$ is not true (see Example 8.3.20).

Remark 8.3.13. Let X be a finite G-space. Then $\overline{y} \leq \overline{x}$ in the quotient space X/G if and only if there exists $g \in G$ such that $y \leq gx$. In particular if X is T_0, so is X/G.

The quotient map $q : X \to X/G$ is open, moreover $q^{-1}(q(U_x)) = \bigcup_{g \in G} gU_x = \bigcup_{g \in G} U_{gx}$. Since $q(U_x) \ni \overline{x}$ is an open set, $U_{\overline{x}} \subseteq q(U_x)$. The other inclusion follows from the continuity of q. Therefore $U_{\overline{x}} = q(U_x)$. Now, $\overline{y} \leq \overline{x}$ if and only if $y \in q^{-1}(U_{\overline{x}}) = \bigcup_{g \in G} U_{gx}$ if and only if there exists $g \in G$ with $y \leq gx$.

Suppose X is T_0, $\overline{y} \leq \overline{x}$ and $\overline{x} \leq \overline{y}$. Then there exist $g, h \in G$ such that $y \leq gx$ and $x \leq hy$. Hence, $y \leq gx \leq ghy$. By Lemma 8.1.1, $y = gx = ghy$ and then $\overline{y} = \overline{x}$.

Proposition 8.3.14. *Let X be a finite T_0-G-space which strongly G-collapses to an invariant subspace Y. Then X/G strongly collapses to Y/G and X^G strongly collapses to Y^G. In particular, if X is contractible, so are X/G and X^G.*

Proof. We can assume there is an elementary strong G-collapse from X to $Y = X \smallsetminus Gx$ where $x \in X$ is a beat point. Suppose $x \in X$ is a down beat point and let $y \prec x$. Then $\overline{y} < \overline{x}$ in X/G and if $\overline{z} < \overline{x}$ there exists g such that $gz < x$. Therefore $gz \leq y$ and $\overline{z} \leq \overline{y}$. This proves that $\overline{x} \in X/G$ is a down beat point and X/G strongly collapses to $X/G \smallsetminus \{\overline{x}\} = Y/G$.

If x is not fixed by G, $Y^G = X^G$. If $x \in X^G$, and $g \in G$, then $gy < gx = x$ and therefore $gy \leq y$. Thus, $gy = y$. This proves that y is also fixed by G and then x is a down beat point of X^G. In particular, $X^G \searrow Y^G$.

If in addition X is contractible, X strongly G-collapses to a G-core which is a point and then X/G and X^G are contractible. $\square$

In fact, the first part of the previous result holds for general spaces. If X is a G-topological space and $Y \subseteq X$ is an equivariant strong deformation retract, Y/G is a strong deformation retract of X/G and so is $Y^G \subseteq X^G$. However if X is a G-topological space which is contractible, X^G need not be contractible. Oliver [64] proved that there are groups which act on disks without fixed points.

Proposition 8.3.15. *Let X be a finite T_0-G-space which G-collapses to Y. Then X^G collapses to Y^G. In particular, if X is G-collapsible, X^G is collapsible.*

Proof. Suppose $X \searrow^{Ge} Y = X \smallsetminus Gx$. If $x \notin X^G$, $Y^G = X^G$. If $x \in X^G$, $\hat{C}_x^X$ is G-invariant and contractible. By Proposition 8.3.14, $\hat{C}_x^{X^G} = (\hat{C}_x^X)^G$ is contractible and then x is a weak point of X^G, which means that $X^G \searrow Y^G$. $\square$

The analogue for quotients is not true. There exist finite T_0-G-spaces such that $X \searrow^G Y$ but X/G does not collapse to Y/G, as the next example shows.

Example 8.3.16. Let X be the following $\mathbb{Z}_2$-space

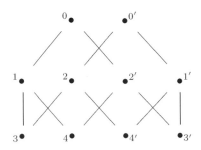

where $\mathbb{Z}_2$ acts by symmetry, $1 \cdot i = i'$ for every $0 \le i \le 4$. Since $0 \in X$ is a weak point, X $\mathbb{Z}_2$-collapses to $Y = X \smallsetminus \{0, 0'\}$. However $X/\mathbb{Z}_2$ does not collapse to $Y/\mathbb{Z}_2$. Moreover, $X/\mathbb{Z}_2$ is contractible while $Y/\mathbb{Z}_2$ is the minimal finite model of the circle.

From Proposition 8.3.15 one easily deduces the next

Corollary 8.3.17. *Let X and Y be equivariantly simple homotopy equivalent finite T_0-G-spaces. Then X^G and Y^G have the same simple homotopy type.*

There is an analogous result of Proposition 8.3.15 for complexes. If K is a G-simplicial complex, K^G denotes the full subcomplex of K spanned by the vertices fixed by the action.

Proposition 8.3.18. *Let K be a finite G-simplicial complex which G-collapses to a subcomplex L. Then K^G collapses to L^G. In particular, if K is G-collapsible, K^G is collapsible.*

Proof. Suppose that $K \searrow^{Ge} L = K \smallsetminus \bigcup_{g \in G} \{g\sigma, g\sigma'\}$, where $\sigma \subsetneq \sigma'$ is a G-collapsible pair. If $\sigma \notin K^G$, $L^G = K^G$. If $\sigma \in K^G$, then $\sigma' \in K^G$, because σ is a free face of σ'. Then $L = K \smallsetminus \{\sigma, \sigma'\}$ and $L^G = K^G \smallsetminus \{\sigma, \sigma'\}$. Since $\sigma \subsetneq \sigma'$ is a collapsible pair in K^G, $K^G \searrow L^G$. $\qquad\square$

Corollary 8.3.19. *If K and L are two finite G-simplicial complexes with the same equivariant simple homotopy type, K^G and L^G have the same simple homotopy type. In particular K has a vertex which is fixed by the action of G if and only if L has a vertex fixed by G.*

Example 8.3.20. Let K be a 1-simplex with the unique nontrivial action of $\mathbb{Z}_2$. The barycentric subdivision K' has a vertex fixed by $\mathbb{Z}_2$ but $K'^{\mathbb{Z}_2} = \emptyset$ therefore K and K' do not have the same equivariant simple homotopy type. On the other hand, $\mathcal{X}(K)$ and $\mathcal{X}(K')$ are contractible, and therefore they have the same equivariant simple homotopy type.

Recall that a map $f : X \to Y$ between finite spaces is called distinguished if $f^{-1}(U_y)$ is contractible for every $y \in Y$. The following result will be used in the next section to compare the equivariant simple homotopy type of $A_p(G)$ and $S_p(G)$.

Proposition 8.3.21. *Let $f : X \to Y$ be a G-map between finite T_0-G-spaces which is distinguished. Then X and Y have the same equivariant simple homotopy type.*

Proof. The non-Hausdorff mapping cylinder $B(f)$ is a G-space with the action induced by X and Y since if $x < y$, then $f(x) \leq y$ and therefore $f(gx) = gf(x) \leq gy$ for every $g \in G$. Moreover, Y is a G-invariant strong deformation retract of $B(f)$ and then $B(f) \searrow^G Y$. On the other hand, $B(f) \searrow^G X$. This follows from the proof of Lemma 4.2.7. Notice that we can remove orbits of minimal points of Y in $B(f)$ and collapse all $B(f)$ into X. □

8.4 Applications to Quillen's Work

Corollary 8.4.1. *$A_p(G)$ and $S_p(G)$ have the same equivariant simple homotopy type.*

Proof. The proof of Proposition 8.2.6 shows that the inclusion $A_p(G) \hookrightarrow S_p(G)$ is a distinguished map. The result then follows by Proposition 8.3.21. □

Corollary 8.4.2. *If G has a nontrivial normal p-subgroup then it has a nontrivial normal elementary abelian p-subgroup.*

Proof. There is a simple algebraic proof of this fact, but we show a shorter one, using the last result. Since $S_p(G) \nearrow^G A_p(G)$, by, Corollary 8.3.17, $S_p(G)^G \searrow A_p(G)^G$. Therefore, if $S_p(G)^G \neq \emptyset$, $A_p(G)^G$ is also nonempty. □

We are now ready to state the result that allows us to study Quillen's conjecture from many different angles.

Theorem 8.4.3. *Let G be a finite group and p a prime integer. The following are equivalent*

1. *G has a nontrivial normal p-subgroup.*
2. *$S_p(G)$ is a contractible finite space.*
3. *$S_p(G)$ is G-collapsible.*
4. *$S_p(G)$ has the equivariant simple homotopy type of a point.*
5. *$\mathcal{K}(S_p(G))$ is G-collapsible.*
6. *$\mathcal{K}(S_p(G))$ has the equivariant simple homotopy type of a point.*
7. *$\mathcal{K}(S_p(G))$ is strong collapsible.*

8. $A_p(G)$ has the equivariant simple homotopy type of a point.
9. $\mathcal{K}(A_p(G))$ has the equivariant simple homotopy type of a point.

Proof. If G has a nontrivial normal p-subgroup, $S_p(G)$ is contractible by Theorem 8.2.2. If $S_p(G)$ is contractible, its G-core is just a point, and since there is a strong G-collapse from a finite T_0-G-space to its G-core, in particular $S_p(G)$ is G-collapsible. If $S_p(G)$ is G-collapsible, $\mathcal{K}(S_p(G))$ is G-collapsible by Theorem 8.3.11 and this implies trivially that $\mathcal{K}(S_p(G))$ has the equivariant simple homotopy type of a point. This, in turn, implies that $S_p(G)$ has the equivariant simple homotopy type of a point by Corollary 8.3.12. If $S_p(G)$ has trivial equivariant simple homotopy type, so does $A_p(G)$ by Corollary 8.4.1, and then so does $\mathcal{K}(A_p(G))$ by Theorem 8.3.11. Now, if $\mathcal{K}(A_p(G))$ has the same equivariant simple homotopy type as a point, then by Corollary 8.3.19 has a vertex which is fixed by the action of G. This vertex corresponds to a nontrivial normal p-subgroup of G. On the other hand, the equivalence between the contractibility of $S_p(G)$ and the strong collapsibility of $\mathcal{K}(S_p(G))$ follows from the main results of Sect. 5.2.   □

On the other hand, recall that the following statements are equivalent: $S_p(G)$ is homotopically trivial, $A_p(G)$ is homotopically trivial, $|\mathcal{K}(S_p(G))|$ is contractible, $|\mathcal{K}(A_p(G))|$ is contractible. So, as a consequence of these equivalences and those given in Theorem 8.4.3, we obtain many different formulations of Quillen's conjecture. The theory exposed in this chapter provides a starting point to attack the conjecture from different angles.

In the following result we will mention one last subspace of $S_p(G)$ that is also weak homotopy equivalent to $S_p(G)$ and $A_p(G)$. In fact it is homotopy equivalent to $S_p(G)$.

Proposition 8.4.4 (Stong). *Let G be a finite group and let p be a prime integer. Let A be the set of nontrivial intersections of Sylow p-subgroups of G. Then A is G-invariant and it is an equivariant strong deformation retract of $S_p(G)$.*

Proof. It is clear that A is G-invariant. Define the retraction $r : S_p(G) \to A$, that assigns to each subgroup $H \subseteq G$, the intersection of all the Sylow p-subgroups containing H. Then r is a continuous map, and $ir \geq 1_{S_p(G)}$. By Proposition 8.3.1, A is an equivariant strong deformation retract of $S_p(G)$.
                                                                                  □

This proof motivates two new constructions that we will introduce in the next chapter, which are used to find the core of some finite spaces called *reduced lattices*. In Sect. 9.2 we will exhibit one last restatement of Quillen's conjecture closely related to the so called Evasiveness conjecture.

Chapter 9
Reduced Lattices

Recall that a poset P is said to be a *lattice* if every two-point set $\{a, b\}$ has a least upper bound $a \vee b$, called *join* or *supremum* of a and b, and a greatest lower bound $a \wedge b$, called *meet* or *infimum*. Any finite lattice has a maximum (and a minimum), and in particular it is a contractible finite space. In this chapter we will study the spaces obtained from a lattice by removing its maximum and its minimum, which are more attractive from a topological point of view. These spaces, here called *reduced lattices*, have been considered before, for instance in [15], where it was proved that if X is a noncomplemented lattice with maximum 1 and minimum 0, then $|\mathcal{K}(X \smallsetminus \{0, 1\})|$ is contractible. We will also introduce the simplicial complex $\mathcal{L}(X)$ associated to any finite space X. We will show that $\mathcal{L}(X)$ has the same weak homotopy type as X when this is a reduced lattice. Connections with strong homotopy types will also be analyzed.

9.1 The Homotopy of Reduced Lattices

Definition 9.1.1. A finite poset X is called a *reduced lattice* if $\hat{X} = D^0 \circledast X \circledast D^0$ is a lattice.

For example, if G is a finite group and p is a prime integer, $S_p(G)$ is a reduced lattice. The finite space $S(G)$ defined in Example 8.1.4 is also a reduced lattice. In contrast, the minimal finite models of the spheres are not.

A subset A of a poset P is *lower bounded* if there exists $x \in P$ such that $x \leq a$ for every $a \in A$. In that case x is called a *lower bound* of A. If the set of lower bounds has a maximum x, we say that x is the *infimum* of A. The notions of *upper bound* and *supremum* are dually defined.

J.A. Barmak, *Algebraic Topology of Finite Topological Spaces and Applications*, Lecture Notes in Mathematics 2032, DOI 10.1007/978-3-642-22003-6_9, © Springer-Verlag Berlin Heidelberg 2011

Proposition 9.1.2. *Let P be a finite poset. The following are equivalent:*

1. *P is a reduced lattice.*
2. *Every lower bounded set of P has an infimum and every upper bounded set has a supremum.*
3. *Every lower bounded set $\{x, y\}$ has infimum.*
4. *Every upper bounded set $\{x, y\}$ has supremum.*

Proof. Straightforward.   □

For instance, the finite space associated to a finite simplicial complex is a reduced lattice. If K is a finite simplicial complex, and $\{\sigma, \sigma'\}$ is lower bounded in $\mathcal{X}(K)$, the simplex $\sigma \cap \sigma'$ is the infimum of $\{\sigma, \sigma'\}$. Moreover, it can be proved that given a finite T_0-space X, there exists a finite simplicial complex K such that $\mathcal{X}(K) = X$ if and only if X is a reduced lattice and every element of X is the supremum of a unique set of minimal elements.

Proposition 9.1.3. *If X is a reduced lattice and $Y \subseteq X$ is a strong deformation retract, Y is also a reduced lattice. In particular, if X is a reduced lattice, so is its core.*

Proof. It suffices to consider the case that $Y = X \smallsetminus \{x\}$, where $x \in X$ is a down beat point. Let $y \prec x$ and let $A = \{a, b\}$ be an upper bounded subset of Y. Then A has a supremum z in X. If x is an upper bound of A in X, $a < x$ and $b < x$ and then $a \le y$, $b \le y$. Therefore $z \neq x$ and then z is the supremum of A in Y. By Proposition 9.1.2, Y is a reduced lattice.   □

However, the fact of being a reduced lattice is not a homotopy type invariant. It is easy to find contractible spaces which are not reduced lattices. Reduced lattices do not describe all homotopy types of finite spaces. For example, since $\mathbb{S}(S^0)$ is minimal and it is not a reduced lattice, no reduced lattice is homotopy equivalent to $\mathbb{S}(S^0)$. On the other hand every finite space X has the weak homotopy type of a reduced lattice, e.g. X'.

The following definition is motivated by Proposition 8.4.4.

Definition 9.1.4. Let X be a reduced lattice. Define the subspace $i(X) \subseteq X$ by $i(X) = \{\inf(A) \mid A$ is a lower bounded subset of maximal elements of $X\}$. Analogously, define $\mathfrak{s}(X) = \{\sup(A) \mid A$ is an upper bounded subset of minimal elements of $X\}$. Here, $\inf(A)$ denotes the infimum of A and $\sup(A)$ its supremum.

Following Stong's proof of Proposition 8.4.4, one can prove that the retraction $r : X \to i(X)$, which sends x to the infimum of the maximal elements of X that are greater than x, is continuous, and that $i(X)$ is a strong deformation retract of X. Similarly, $\mathfrak{s}(X) \subseteq X$ is a strong deformation retract.

Example 9.1.5. Let $n \geq 2$ and let P_n be the poset of proper positive divisors of n with the order given by: $a \leq b$ if a divides b. If n is square free, P_n is homeomorphic to $\mathcal{X}(\dot{\sigma})$ where σ is a $(k-1)$-simplex, k being the number of primes dividing n. In fact, if $p_1, p_2, \ldots, p_k$ are the prime divisors of n, and $\sigma = \{p_1, p_2, \ldots, p_k\}$ is a simplex, the map $f : P_n \to \mathcal{X}(\dot{\sigma})$ defined by $f(d) = \{p_i \mid p_i \text{ divides } d\}$, is a homeomorphism. In particular, $|\mathcal{K}(P_n)| = |(\dot{\sigma})'|$ is homeomorphic to the $(k-2)$-dimensional sphere.

If n is not square free, we show that P_n is contractible. Note that P_n is a reduced lattice with the infimum given by the greatest common divisor. Since n is not square free, the product of the prime divisors of n is a proper divisor of n and it is the maximum of $\mathfrak{s}(P_n)$. Thus, $\mathfrak{s}(P_n)$ is contractible and then, so is P_n.

Proposition 9.1.6. *Let X be a reduced lattice. The following are equivalent*

1. X is a minimal finite space.
2. $\mathfrak{i}(X) = \mathfrak{s}(X) = X$.

Proof. If X is minimal, the unique strong deformation retract of X is X itself. Therefore $\mathfrak{i}(X) = \mathfrak{s}(X) = X$. Conversely, suppose this equality holds and that $x \in X$ is a down beat point with $y \prec x$. Since $x \in X = \mathfrak{s}(X)$, x is the supremum of a set M of minimal elements of X. Since x is not minimal, every element of M is strictly smaller than x, and therefore y is an upper bound of M. This contradicts the fact that $x = \sup(M)$. Then X does not have down beat points and analogously it does not have up beat points, so it is minimal.   □

If X is a reduced lattice, $\mathfrak{i}(X)$ is a strong deformation retract of X, which is a reduced lattice by Proposition 9.1.3. Analogously $\mathfrak{s}(\mathfrak{i}(X))$ is a strong deformation retract of X and it is a reduced lattice. The sequence

$$X \supseteq \mathfrak{i}(X) \supseteq \mathfrak{s}\mathfrak{i}(X) \supseteq \mathfrak{i}\mathfrak{s}\mathfrak{i}(X) \supseteq \ldots$$

is well defined and it stabilizes in a space Y which is a strong deformation retract of X and a minimal finite space by Proposition 9.1.6. Therefore, in order to obtain the core of a reduced lattice, one can carry out alternatively the constructions $\mathfrak{i}$ and $\mathfrak{s}$, starting from any of them.

Example 9.1.7. Let K be the simplicial complex which consists of two 2-simplices with a common 1-face. Since K is strong collapsible, so is $\mathcal{X}(K)$. Another way to see this is the following: $\mathcal{X}(K)$ is a reduced lattice with two maximal elements, $\mathfrak{i}(\mathcal{X}(K))$ has just three points, and $\mathfrak{s}\mathfrak{i}(\mathcal{X}(K))$ is the singleton.

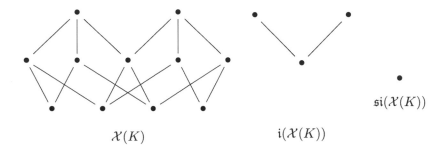

$$\mathcal{X}(K) \qquad\qquad \mathrm{i}(\mathcal{X}(K)) \qquad\qquad \mathfrak{si}(\mathcal{X}(K))$$

Although there are many reduced lattices which are minimal finite spaces, a reduced lattice X is a minimal finite model if and only if it is discrete. If X is not discrete, there is a point $x \in X$ which is not minimal and we can apply Osaki's open reduction (Theorem 6.1.1) to obtain a smaller model X/U_x.

Given a finite T_0-space X, we will denote by $\mathrm{Max}(X)$ the set of maximal elements of X.

Proposition 9.1.8. *If X and Y are two finite T_0-spaces and $f : X \to Y$ is a continuous map, there exists $g : X \to Y$, $g \ge f$, such that $g(\mathrm{Max}(X)) \subseteq \mathrm{Max}(Y)$.*

Proof. Let $g \ge f$ be a maximal element in Y^X. Suppose there exists $x \in \mathrm{Max}(X)$ such that $g(x) \notin \mathrm{Max}(Y)$. Then, there exists $y > g(x)$, and the map $\widetilde{g} : X \to Y$ which coincides with g in $X \smallsetminus \{x\}$ and such that $\widetilde{g}(x) = y$ is continuous and $\widetilde{g} > g$, which is a contradiction. Therefore $g(\mathrm{Max}(X)) \subseteq \mathrm{Max}(Y)$.   □

Remark 9.1.9. Let X be a finite T_0-space and Y a reduced lattice. If $f, g : X \to Y$ are two maps which coincide in $\mathrm{Max}(X)$, then $f \simeq g$. Define $h : X \to Y$ by $h(x) = \inf(\{f(x') \mid x' \in \mathrm{Max}(X) \text{ and } x' \ge x\})$. Clearly h is continuous and $h \ge f$. Analogously $h \ge g$ and then $f \simeq g$.

We deduce then from Proposition 9.1.8 that if X is a finite T_0-space and Y is a reduced lattice, $\#[X, Y] \le (\#\mathrm{Max}(Y))^{\#\mathrm{Max}(X)}$, where $[X, Y]$ denotes the set of homotopy classes of maps $X \to Y$.

9.2   The $\mathcal{L}$ Complex

Let X be a finite T_0-space. The simplicial complex $\mathcal{L}(X)$ is defined as follows. Its vertices are the maximal elements of X and its simplices are the subsets of $\mathrm{Max}(X)$ which are lower bounded. In other words, $\mathcal{L}(X)$ is the nerve of the cover $\mathcal{U}_0$ of X given by the minimal open sets of the maximal elements.

In general, X is not a finite model of $\mathcal{L}(X)$. If $X = \mathbb{S}S^0$ is the minimal finite model of S^1, $\mathcal{L}(X)$ is a 1-dimensional simplex. However, if X is a reduced lattice, then each intersection of minimal open sets has a maximum or it is empty. In particular, intersections of elements of $\mathcal{U}_0$ are empty or homotopically trivial. By Theorem 5.3.1, $|\mathcal{L}(X)|$ and X are weak homotopy equivalent. An alternative proof of the fact that $|\mathcal{L}(X)|$ and $|\mathcal{K}(X)|$ are homotopy equivalent is to apply the so called Crosscut Theorem [16, Theorem 10.8].

Notice that if K is a finite simplicial complex, $\mathcal{L}(\mathcal{X}(K))$ is exactly the nerve $\mathcal{N}(K)$ of K. On the other hand $\mathcal{L}(\mathcal{X}(K)^{op}) = K$.

Two reduced lattices X and Y have the same weak homotopy type if and only if $\mathcal{L}(X)$ and $\mathcal{L}(Y)$ are homotopy equivalent. The following results study the relationship between homotopy types of finite spaces and strong homotopy types of the $\mathcal{L}$ complexes.

In Proposition 9.1.8, we observed that given any map $f : X \to Y$ between finite T_0-spaces, there exists $g \geq f$ such that $g(\mathrm{Max}(X)) \subseteq \mathrm{Max}(Y)$. A map $f : X \to Y$ satisfying $f(\mathrm{Max}(X)) \subseteq \mathrm{Max}(Y)$ will be called *good*. Note that a good map $f : X \to Y$ between finite T_0-spaces induces a simplicial map $\mathcal{L}(f) : \mathcal{L}(X) \to \mathcal{L}(Y)$ given by $\mathcal{L}(f)(x) = f(x)$ since f maps a lower bound of $A \subseteq X$ to a lower bound of $f(A) \subseteq Y$.

Theorem 9.2.1. *If X and Y are homotopy equivalent finite T_0-spaces, $\mathcal{L}(X)$ and $\mathcal{L}(Y)$ have the same strong homotopy type.*

Proof. We can assume that there is a homotopy equivalence $f : X \to Y$ which is a good map. Let $g : Y \to X$ be a homotopy inverse of f and which is also good. We will prove that the composition $\mathcal{L}(g)\mathcal{L}(f) : \mathcal{L}(X) \to \mathcal{L}(Y)$ lies in the same contiguity class as the identity $1_{\mathcal{L}(X)}$. Symmetrically, $\mathcal{L}(f)\mathcal{L}(g) \sim 1_{\mathcal{L}(Y)}$ and then $\mathcal{L}(f)$ is a strong equivalence. Thus, by Corollary 5.1.11, $\mathcal{L}(X)$ and $\mathcal{L}(Y)$ are strong homotopy equivalent.

Since $gf \simeq 1_X$, there exists a fence $gf = h_0 \geq h_1 \leq h_2 \geq h_3 \leq \ldots \geq h_{2k-1} \leq h_{2k} = 1_X$. Moreover, by Proposition 9.1.8 we can assume that h_{2i} is good for every $0 \leq i \leq k$. It suffices to show then that $\mathcal{L}(h_{2i})$ and $\mathcal{L}(h_{2i+2})$ are contiguous for each $0 \leq i < k$. Let σ be a simplex of $\mathcal{L}(X)$. Then, there exists a lower bound $x \in X$ for all the points of σ. Then $h_{2i+1}(x) \leq h_{2i}(x)$ is a lower bound of $h_{2i}(\sigma)$. Analogously, $h_{2i+1}(x)$ is a lower bound of $h_{2i+2}(\sigma)$. Since $h_{2i}(\sigma) \cup h_{2i+2}(\sigma)$ has a lower bound, it is a simplex of $\mathcal{L}(X)$. This proves that $\mathcal{L}(h_{2i})$ and $\mathcal{L}(h_{2i+2})$ are contiguous. $\square$

Corollary 9.2.2. *Let K and L be two finite simplicial complexes. If $\mathcal{X}(K) \simeq \mathcal{X}(L)$, then K and L have the same strong homotopy type.*

Proof. If $\mathcal{X}(K) \simeq \mathcal{X}(L)$, $\mathcal{X}(K)^{op} \simeq \mathcal{X}(L)^{op}$ by Corollary 1.2.7, and by Theorem 9.2.1 $K = \mathcal{L}(\mathcal{X}(K)^{op})$ and $L = \mathcal{L}(\mathcal{X}(L)^{op})$ have the same strong homotopy type. $\square$

In Corollary 5.2.7 it was proved that the contractibility of X' implies the contractibility of X. That proof uses Theorem 5.2.6. We give an alternative proof here using the last corollary. Suppose that X' is contractible. Let Y be a the core of X. Since Y' is also contractible, $\mathcal{X}(\mathcal{K}(Y)) = Y' \simeq \mathcal{X}(*)$. By Corollary 9.2.2, $\mathcal{K}(Y)$ is strong collapsible. However, by Proposition 5.2.5, $\mathcal{K}(Y)$ is a minimal complex and therefore $Y = *$. Hence, X is contractible.

The converse of Theorem 9.2.1 is true when X and Y are reduced lattices. First we prove the following

Lemma 9.2.3. *Let X be a reduced lattice. Then $\mathcal{X}(\mathcal{L}(X)) \simeq X^{op}$.*

Proof. Define $f : \mathcal{X}(\mathcal{L}(X)) \to X^{op}$ by $f(\sigma) = \inf(\sigma)$, where $\inf(\sigma)$ is the infimum of σ in X. Let $g : X^{op} \to \mathcal{X}(\mathcal{L}(X))$ be defined by $g(x) = \{y \in \mathrm{Max}(X) \mid y \geq_X x\}$. Clearly f and g are order preserving. Moreover $gf(\sigma) = \{y \in \mathrm{Max}(X) \mid y \geq_X \inf(\sigma)\} \supseteq \sigma$. Then $gf \geq 1_{\mathcal{X}(\mathcal{L}(X))}$. On the other hand $fg(x) = \inf(\{y \in \mathrm{Max}(X) \mid y \geq_X x\}) \geq_X x$. Then $fg \leq_{X^{op}} 1_{X^{op}}$. Thus, f is a homotopy equivalence.   □

Theorem 9.2.4. *Let X and Y be two reduced lattices. Then $X \simeq Y$ if and only if $\mathcal{L}(X)$ and $\mathcal{L}(Y)$ have the same strong homotopy type.*

Proof. By Theorem 9.2.1 it only remains to prove one implication. Suppose that $\mathcal{L}(X)$ and $\mathcal{L}(Y)$ have the same strong homotopy type. By Theorem 5.2.1, $\mathcal{X}(\mathcal{L}(X)) \simeq \mathcal{X}(\mathcal{L}(Y))$ and by Lemma 9.2.3, $X^{op} \simeq Y^{op}$. Then by Corollary 1.2.7, X and Y have the same homotopy type.   □

The $\mathcal{L}$ construction can be used to give a new restatement of Quillen's conjecture, different from those mentioned in Chap. 8.

Definition 9.2.5. Let G be a finite group and p a prime integer dividing the order of G. The complex $L_p(G) = \mathcal{L}(S_p(G))$ is the complex whose vertices are the Sylow p-subgroups of G and whose simplices are sets of Sylow p-subgroups with nontrivial intersection.

Since $S_p(G)$ is a reduced lattice, $|L_p(G)|$ is weak homotopy equivalent to $S_p(G)$. Any normal p-subgroup of G is contained in the intersection of all the Sylow p-subgroups. Conversely, the intersection of the Sylow p-subgroups is a normal subgroup of G. Then, G has a nontrivial normal p-subgroup if and only if the intersection of all the Sylow p-subgroups is nontrivial, or equivalently if $L_p(G)$ is a simplex. Therefore, Quillen's conjecture can be restated as follows:

Restatement of Quillen's conjecture: if $|L_p(G)|$ is contractible, it is a simplex.

The complex $L_p(G)$ is what is called a *vertex homogeneous* simplicial complex. This means that the automorphism group of $L_p(G)$ acts transitively on the vertices. In other words, given any two vertices $v, w \in L_p(G)$, there exists a simplicial automorphism $\varphi : L_p(G) \rightarrow L_p(G)$ such that $\varphi(v) = w$. The reason is that any two Sylow p-subgroups are conjugate and that any automorphism of $S_p(G)$ induces an automorphism of $L_p(G)$. Therefore, the conjecture claims that a particular contractible vertex homogeneous complex must be a simplex. In general this is not true. In [47] examples are shown of contractible vertex homogeneous complexes which are not simplices. This problem is related to the so called Evasiveness conjecture [46]. In [11] it is proved that a vertex homogeneous complex which is strong collapsible must be a simplex. This is used to reduce the Evasiveness conjecture to the case of the minimal simplicial complexes.

Chapter 10
Fixed Points and the Lefschetz Number

In Chap. 8 we studied fixed point sets of group actions. Now we turn our
attention to fixed point sets of continuous maps between finite spaces and
their relationship with the fixed point sets of the associated simplicial maps.
We analyze well-known results on the fixed point theory of finite posets from
the perspective of finite spaces. An extensive treatment of the fixed point
theory for posets appears in Baclawski and Björner's paper [4]. We use the
poset version of the Lefschetz Theorem to analyze fixed points of simplicial
automorphisms, providing an alternative approach to Oliver's work [63, 64].

10.1   The Fixed Point Property for Finite Spaces

If X is a topological space and $f : X \to X$ is a continuous map, we denote
by $X^f = \{x \in X \mid f(x) = x\}$ the set of fixed points of f. For a simplicial
map $\varphi : K \to K$, K^φ denotes the full subcomplex spanned by the vertices
fixed by φ.

We say that a topological space X has the *fixed point property* if any map
$f : X \to X$ has a fixed point. For instance, any disk has the fixed point
property by the Brouwer fixed-point Theorem.

The proof of the following result uses the construction $f^\infty(X)$ studied in
Sect. 3.4 (cf. [26, Proposition 1]).

Proposition 10.1.1. *A finite T_0-space X has the fixed point property if and
only if all its retracts have the fixed point property with respect to automorphisms.*

Proof. The first implication holds in general, if X is a topological space with
the fixed point property, every retract of X also has that property. Conversely,
if $f : X \to X$ is a continuous map, then $f^\infty(X)$ is a retract of X and

J.A. Barmak, *Algebraic Topology of Finite Topological Spaces and*   129
Applications, Lecture Notes in Mathematics 2032,
DOI 10.1007/978-3-642-22003-6_10, © Springer-Verlag Berlin Heidelberg 2011

$f|_{f^\times(X)} : f^\infty(X) \to f^\infty(X)$ is an automorphism. If $f|_{f^\times(X)}$ has a fixed point, so does f. $\square$

The following is a well-known result:

Proposition 10.1.2. *Let X be a finite T_0-space, and let $f, g : X \to X$ be two homotopic maps. Then f has a fixed point if and only if g has a fixed point.*

Proof. Without loss of generality, we can assume that $g \leq f$. If $f(x) = x$, $g(x) \leq f(x) = x$. Then $g^{i+1}(x) \leq g^i(x)$ for every $i \geq 0$ and then there exists i such that $g^{i+1}(x) = g^i(x)$. Therefore, $g^i(x)$ is a fixed point of g. $\square$

We use last proposition to prove a generalization of the fact that retracts of finite T_0-spaces with the fixed point property also have that property.

Proposition 10.1.3. *Let X and Y be finite T_0-spaces such that there exist continuous maps $f : X \to Y$ and $g : Y \to X$ with $fg \simeq 1_Y$. Then if X has the fixed point property, so does Y.*

Proof. Let $h : Y \to Y$ be a continuous map. Then map $ghf : X \to X$ fixes some point $x \in X$. Therefore $f(x) \in Y$ is a fixed point of $fgh : Y \to Y$. Since $h \simeq fgh$, h has a fixed point by Proposition 10.1.2. $\square$

Immediately one deduces the following

Corollary 10.1.4. *The fixed point property is a homotopy type invariant of finite T_0-spaces.*

A different proof of this result appears for example in [71, Proposition 1] and in [81, Corollary 3.16].

Note the T_0 hypothesis in the last three results is necessary. If X is an indiscrete space of two points, then X is contractible but it does not have the fixed point property. Both homeomorphisms $X \to X$ are homotopic but one has fixed points while the other does not. The fixed point property is not a homotopy invariant for non-finite spaces either, not even if we restrict to the class of compact metric spaces. There are examples of contractible compact metric spaces without the fixed point property (see [42]).

If X is a contractible finite T_0-space, then it has the fixed point property by Corollary 10.1.4. In [71] Rival observed that the converse is not true. The example provided is the space X of Example 4.3.3. In [71] no method is suggested for proving that such a space X has the fixed point property. However we know that X is collapsible and in particular homotopically trivial. We will now recall the Lefschetz fixed point Theorem for compact polyhedra and the version of this result for posets. Any of these results can be applied directly to prove that every homotopically trivial finite T_0-space has the fixed point property.

Let M be a finitely generated $\mathbb{Z}$-module, and $T(M)$ its torsion submodule. An endomorphism $\varphi : M \to M$ induces a morphism $\overline{\varphi} : M/T(M) \to M/T(M)$ between finite-rank free $\mathbb{Z}$-modules. The *trace* $tr(\varphi)$ of φ is the trace of $\overline{\varphi}$. Namely, if $\{e_1, e_2, \ldots, e_r\}$ is a basis of $M/T(M)$ and $\overline{\varphi}(e_i) = \sum_j m_{ij} e_j$, $tr(\varphi) = \sum\limits_{i=1}^{r} m_{ii}$. If K is a compact polyhedron, $H_*(K)$ is a finitely generated graded abelian group, that is $H_n(K)$ is finitely generated for every $n \geq 0$ and is nontrivial only for finitely many n. If $f : K \to K$ is a continuous map, the *Lefschetz number* of f is defined by

$$\lambda(f) = \sum_{n \geq 0} (-1)^n tr(f_n), \tag{10.1}$$

where $f_n : H_n(K) \to H_n(K)$ are the induced morphisms in homology.

Notice that the Lefschetz number of the identity $1_K : K \to K$ coincides with the Euler characteristic of K.

Theorem 10.1.5 (Lefschetz Theorem). *Let K be a compact polyhedron and let $f : K \to K$ be a continuous map. Then, if $\lambda(f) \neq 0$, f has a fixed point.*

In particular, if K is contractible, $\lambda(f) = 1$ for every map $f : K \to K$ and then f has a fixed point. This generalizes the well-known Brouwer fixed-point Theorem for disks. A proof of the Lefschetz Theorem can be found in [75, Theorem 4.7.7].

If X is a finite T_0-space, its homology is finitely generated as well and therefore we can define the Lefschetz number $\lambda(f)$ of a map $f : X \to X$ as in 10.1. Note that $\lambda(f) = \lambda(|\mathcal{K}(f)|)$ by Remark 1.4.7.

The version of this theorem for finite spaces is the following. For details on the proof we refer the reader to [4, Theorem 1.1] (see also [59]).

Theorem 10.1.6 (Baclawski-Björner). *Let X be a finite T_0-space and $f : X \to X$ a continuous map. Then $\lambda(f) = \chi(X^f)$. In particular, if $\lambda(f) \neq 0$, $X^f \neq \emptyset$.*

Corollary 10.1.7. *Any homotopically trivial finite T_0-space has the fixed point property.*

Moreover, finite T_0-spaces with trivial rational homology groups also have the fixed point property. We proved in Corollary 10.1.4 that the fixed point property is a homotopy invariant. The following example shows that it is not a weak homotopy invariant and, at the same time, that the hypothesis of having trivial rational homology is not needed to have the fixed point property.

Example 10.1.8 (Baclawski-Björner). The fixed point property is not a weak homotopy invariant, nor a simple homotopy invariant. In [4, Example 2.4], Baclawski and Björner considered the regular CW-complex K which is the boundary of a pyramid with square base (see Fig. 10.1).

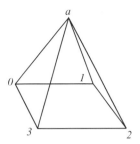

Fig. 10.1 K

The associated finite space $X = \mathcal{X}(K)$ is a finite model of S^2. Let $f :$ $X \to X$ be a continuous map. If f is onto, it is an automorphism and then the vertex of the top of the pyramid is fixed by f since it is the unique point covered by 4 points. If f is not onto, $\mathcal{K}(f) : S^2 \to S^2$ is not onto and then $\mathcal{K}(f)$ is nullhomotopic. Therefore $\lambda(f) = \lambda(|\mathcal{K}(f)|) = 1$ and then $X^f \neq \emptyset$.

On the other hand, the minimal finite model of S^2

is simple homotopy equivalent to X and does not have the fixed point property since the symmetry is fixed point free.

With a similar idea as in this example it is possible to construct finite models of each sphere S^n, $n \geq 2$, having the fixed point property.

Open problem: Which spaces have finite models with the fixed point property?

A simple case to start studying this question seems to be the one-dimensional sphere.

Proposition 10.1.9. *Let X be a finite T_0-space. If X' has the fixed point property, then so does X.*

Proof. If $f : X \to X$ is continuous and X' has the fixed point property, f leaves fixed a nonempty chain of X and hence, all its points.   □

Remark 10.1.10. The converse of Proposition 10.1.9 is not true. Consider the regular CW-complex K of Example 10.1.8. Denote by $0, 1, 2, 3 \in \mathbb{Z}_4$ the vertices in the base of the pyramid K and by a the fifth vertex, as shown in Fig. 10.1. A cell e of K will be identified with the set of vertices in $\bar{e}$. Define $f : \mathcal{X}(K) \to \mathcal{X}(K)^{op}$ as follows: $f(\{a\}) = \{0, 1, 2, 3\}$, $f(\{0, 1, 2, 3\}) = \{a\}$ and for every $n \in \mathbb{Z}_4$, $f(\{n\}) = \{n, n+1, a\}$, $f(\{n, a\}) = \{n, n+1\}$, $f(\{n, n+1\}) = \{n+1, a\}$, $f(\{n, n+1, a\}) = \{n+1\}$. It is easy to see that f is order preserving and that $f' : \mathcal{X}(K)' \to \mathcal{X}(K)'$ does not have fixed points. However, as it was shown in Example 10.1.8, $\mathcal{X}(K)$ has the fixed point property.

10.2   On the Lefschetz Theorem for Simplicial Automorphisms

Proposition 10.2.1. *Let $\varphi : K \to K$ be a simplicial automorphism. Then $|K|^{|\varphi|} = |(K')^{\varphi'}|$.*

Proof. Let $x \in |K'| = |K|$. Then $x = \sum \alpha_i b(\sigma_i)$ is a convex combination of the barycenters of simplices $\sigma_0 \subsetneq \sigma_1 \subsetneq \ldots \subsetneq \sigma_k$ of K with $\alpha_i > 0$ for every i. Suppose $x \in |(K')^{\varphi'}|$. This is equivalent to saying that each of the vertices $b(\sigma_i)$ in the support of x is fixed by φ', or, in other words, that $\varphi(\sigma_i) = \sigma_i$ for every i. If we now consider $x \in |K|$,

$$x = \sum \alpha_i \sum_{v \in \sigma_i} \frac{v}{\#\sigma_i}$$

and

$$|\varphi|(x) = \sum \alpha_i \sum_{v \in \sigma_i} \frac{\varphi(v)}{\#\sigma_i}.$$

Since $\varphi(\sigma_i) = \sigma_i$, $\sum_{v \in \sigma_i} \frac{\varphi(v)}{\#\sigma_i} = \sum_{v \in \sigma_i} \frac{v}{\#\sigma_i}$, and then $|\varphi|(x) = x$. This proves one inclusion.

Before proving the other inclusion, note that if $v \in \sigma_i \setminus \sigma_{i-1}$, the coordinate of v in $x = \sum \alpha_i \sum_{v \in \sigma_i} \frac{v}{\#\sigma_i}$ is

$$\frac{\alpha_i}{\#\sigma_i} + \frac{\alpha_{i+1}}{\#\sigma_{i+1}} + \ldots + \frac{\alpha_k}{\#\sigma_k}.$$

Since φ is an isomorphism, the coordinate of $\varphi(v)$ in $|\varphi|(x) = \sum \alpha_i \sum_{v \in \sigma_i} \frac{\varphi(v)}{\#\sigma_i}$ is also $\frac{\alpha_i}{\#\sigma_i} + \frac{\alpha_{i+1}}{\#\sigma_{i+1}} + \ldots + \frac{\alpha_k}{\#\sigma_k}$.

Suppose now that $x \in |K|^{|\varphi|}$. In this case, $\varphi(v) \in support(|\varphi|(x)) = support(x)$ and therefore $\varphi(v) \in \sigma_k$. If $\varphi(v) \in \sigma_j \smallsetminus \sigma_{j-1}$, the coordinate of $\varphi(v)$ in x is $\frac{\alpha_j}{\#\sigma_j} + \frac{\alpha_{j+1}}{\#\sigma_{j+1}} + \ldots + \frac{\alpha_k}{\#\sigma_k}$. Since $|\varphi|(x) = x$,

$$\frac{\alpha_i}{\#\sigma_i} + \frac{\alpha_{i+1}}{\#\sigma_{i+1}} + \ldots + \frac{\alpha_k}{\#\sigma_k} = \frac{\alpha_j}{\#\sigma_j} + \frac{\alpha_{j+1}}{\#\sigma_{j+1}} + \ldots + \frac{\alpha_k}{\#\sigma_k}$$

and then $i = j$. This proves that $\varphi(\sigma_i \smallsetminus \sigma_{i-1}) \subseteq \sigma_i \smallsetminus \sigma_{i-1}$. Thus, $\varphi(\sigma_i) \subseteq \sigma_i$ and then $\varphi(\sigma_i) = \sigma_i$ for every i. Therefore $x \in |(K')^{\varphi'}|$, which proves the other inclusion.   □

Since $X^f \subseteq X$, $\mathcal{K}(X^f)$ is the full subcomplex of $\mathcal{K}(X)$ spanned by the vertices fixed by f. By definition, this subcomplex is $\mathcal{K}(X)^{\mathcal{K}(f)}$. Therefore we have,

Remark 10.2.2. Let X be a finite T_0-space and let $f : X \to X$ be a continuous map. Then $\mathcal{K}(X^f) = \mathcal{K}(X)^{\mathcal{K}(f)}$.

Corollary 10.2.3. Let K be a finite simplicial complex and $\varphi : K \to K$ a simplicial automorphism. Then $\mathcal{X}(K)^{\mathcal{X}(\varphi)}$ is a finite model of $|K|^{|\varphi|}$.

Proof. By Proposition 10.2.1, $|K|^{|\varphi|} = |(K')^{\varphi'}| = |\mathcal{K}(\mathcal{X}(K))^{\mathcal{K}(\mathcal{X}(\varphi))}|$ and by Remark 10.2.2, this coincides with $|\mathcal{K}(\mathcal{X}(K)^{\mathcal{X}(\varphi)})|$ which is weak homotopy equivalent to $\mathcal{X}(K)^{\mathcal{X}(\varphi)}$.   □

The following is a stronger version of Lefschetz Theorem 10.1.5 for simplicial automorphisms. A different proof can be found in [63] (see also [39, Theorem 1.8]).

Corollary 10.2.4. Let K be a finite simplicial complex and let $\varphi : K \to K$ be a simplicial automorphism. Then $\chi(|K|^{|\varphi|}) = \lambda(|\varphi|)$.

Proof. The diagram

$$
\begin{array}{ccc}
 & |\varphi| & \\
|K| & \longrightarrow & |K| \\
\Big\downarrow{\scriptstyle\mu_K} & & \Big\downarrow{\scriptstyle\mu_K} \\
\mathcal{X}(K) & \xrightarrow{\ \mathcal{X}(\varphi)\ } & \mathcal{X}(K)
\end{array}
$$

commutes up to homotopy and $(\mu_{K*})_n : H_n(|K|) \to H_n(\mathcal{X}(K))$ is an isomorphism for every $n \geq 0$. Then $|\varphi|_* = (\mu_{K*})^{-1}\mathcal{X}(\varphi)_*\mu_{K*} : H_n(|K|) \to H_n(|K|)$ and $tr((|\varphi|_*)_n) = tr((\mathcal{X}(\varphi)_*)_n)$. Therefore $\lambda(|\varphi|) = \lambda(\mathcal{X}(\varphi))$. By Corollary 10.2.3 and the finite space version of the Lefschetz Theorem 10.1.6, $\chi(|K|^{|\varphi|}) = \chi(\mathcal{X}(K)^{\mathcal{X}(\varphi)}) = \lambda(\mathcal{X}(\varphi)) = \lambda(|\varphi|)$.   □

From this corollary we obtain an alternative proof of a result of R. Oliver [64, Lemma 1].

Proposition 10.2.5 (Oliver). *Assume that $\mathbb{Z}_n$ acts on a $\mathbb{Q}$-acyclic finite simplicial complex K. Then $\chi(|K|^{\mathbb{Z}_n}) = 1$.*

Proof. Let g be a generator of $\mathbb{Z}_n$ and $\varphi : K \to K$ the multiplication by g. Then $\chi(|K|^{\mathbb{Z}_n}) = \chi(|K|^{|\varphi|}) = \lambda(|\varphi|) = 1$, since K is $\mathbb{Q}$-acyclic. $\qquad\square$

We conclude this chapter with a result on the trace of the morphism induced by a map between finite spaces.

Suppose that X is a finite model of the circle and that $f : X \to X$ is a map. Then $f_* : H_1(X) \to H_1(X)$ is a map $\mathbb{Z} \to \mathbb{Z}$. However, the only possible morphisms that can appear in this way are $0, 1_{\mathbb{Z}}$ and $-1_{\mathbb{Z}}$. We prove this and a more general fact in the following result.

Proposition 10.2.6. *Let $f : X \to X$ be an endomorphism of a finite T_0-space X and let $n \geq 0$. Let $f_n : H_n(X; \mathbb{Q}) \to H_n(X; \mathbb{Q})$ be the induced map in homology. If $\dim_{\mathbb{Q}} H_n(X; \mathbb{Q}) = r$, f_n is a matrix of order r with rational entries well defined up to similarity. Suppose that $\lambda \in \mathbb{C}$ is an eigenvalue of f_n considered as a complex matrix. Then $\lambda = 0$ or λ is a root of unity.*

Proof. Since X is finite, there exist $s \neq t \in \mathbb{N}$ such that $f^s = f^t$. Then $f_n^s = f_n^t$ and $\lambda^s = \lambda^t$. $\qquad\square$

Corollary 10.2.7. *Under the hypothesis of the previous proposition, $-r \leq tr(f_n) \leq r$. In particular, since f_n has integer entries, $tr(f_n) \in \{-r, -r + 1, \ldots, r - 1, r\}$.*

Chapter 11
The Andrews–Curtis Conjecture

The Poincaré conjecture is one of the most important problems in the history of Mathematics. The generalized versions of the conjecture for dimensions greater than 3 were proved between 1961 and 1982 by Smale, Stallings, Zeeman and Freedman. However, the original problem remained open for a century until Perelman finally proved it some years ago [66–68]. His proof is based on Hamilton's theory of Ricci flow. An alternative combinatorial proof of the Poincaré conjecture would be a great achievement.

The Zeeman conjecture and the Andrews–Curtis conjecture are closely related to the original Poincaré conjecture. Moreover, with the proof of the Poincaré conjecture it is now known that both conjectures are true for a class of complexes called *standard spines*. However both conjectures are still open.

In this chapter we will introduce the class of *quasi constructible complexes* which are built recursively by attaching smaller quasi constructible complexes. Using techniques of finite spaces we will prove that contractible quasi constructible complexes satisfy the Andrews–Curtis conjecture. Quasi constructible complexes generalize the notion of *constructible complexes* which was deeply studied by Hachimori in [33].

11.1 n-Deformations and Statements of the Conjectures

The Poincaré conjecture, originally formulated in 1904, can be stated as follows:

Poincaré conjecture: Every simply connected closed 3-manifold is homeomorphic to S^3.

Zeeman proved the 5-dimensional version of the Poincaré conjecture, but he also studied the original problem. In [87] he showed that, although no triangulation of the Dunce Hat D is collapsible, the cylinder $D \times I$ is

J.A. Barmak, *Algebraic Topology of Finite Topological Spaces and Applications*, Lecture Notes in Mathematics 2032,
DOI 10.1007/978-3-642-22003-6_11, © Springer-Verlag Berlin Heidelberg 2011

polyhedrally collapsible. A polyhedral collapse is somewhat less rigid than a simplicial collapse and is more suitable when working in the category of polyhedra and piecewise linear maps. An elementary polyhedral collapse from a polyhedron K to a subpolyhedron L consists of the removal of an n-ball B^n of K which intersects L in an $(n-1)$-ball contained in the boundary of B^n. If there is a polyhedral collapse from K to L, then there exist triangulations K_0, L_0 of K and L such that $K_0 \searrow L_0$. For more details see [31, 50, 87]. Zeeman conjectured that this property holds more generally for any contractible 2-complex:

Zeeman conjecture: If K is a contractible compact polyhedron, then $K \times I$ is polyhedrally collapsible.

Zeeman proved in [87] that his conjecture implies the Poincaré conjecture. In [30] Gillman and Rolfsen proved that the Poincaré conjecture is equivalent to the Zeeman conjecture when restricted to *standard spines* (see also [72]). Zeeman's conjecture is still not proved nor disproved. In these notes we will not work with the polyhedral version of collapse.

A *balanced presentation* of a group G is a presentation $\langle x_1, x_2, \ldots, x_n \mid r_1, r_2, \ldots r_n \rangle$ with the same number of generators than relators. Given a presentation $\langle x_1, x_2, \ldots, x_n \mid r_1, r_2, \ldots r_m \rangle$ of G, we consider the following four operations which modify this one to obtain a new presentation of G.

(1) Replace a relator r_i by r_i^{-1}.
(2) Replace a relator r_i by $r_i r_j$ where $j \neq i$.
(3) Replace a relator r_i by $g r_i g^{-1}$ where g is any element in the free group generated by $x_1, x_2, \ldots, x_n$.
(4) Add a generator x_{n+1} and a relator $r_{m+1} = x_{n+1}$.

Andrews–Curtis conjecture: Any balanced presentation of the trivial group can be transformed into the trivial presentation by performing repeatedly the operations (1)–(4) and the inverse of operation (4).

We are particularly interested in a topological version of this conjecture.

Definition 11.1.1. Let $n \geq 1$. We say that a complex K *n-deforms* to another complex L if we can obtain L from K by a sequence of collapses and expansions in such a way that all the complexes involved in the deformation have dimension less than or equal to n.

Geometric Andrews–Curtis conjecture: Any contractible compact 2-polyhedron 3-deforms to a point.

The Geometric Andrews–Curtis conjecture is equivalent to the Andrews–Curtis conjecture. It is clear that Zeeman's conjecture implies the Andrews–Curtis conjecture. However, the later is also an open problem. For more references see [2, 50, 72].

The analogous version of the Andrews–Curtis conjecture for higher dimensions is known to be true. More specifically, Wall proved in [82, Theorem 1] the following result which is an improvement of a result of Whitehead.

Theorem 11.1.2 (Whitehead-Wall). *Let $n \geq 3$. If K and L are simple homotopy equivalent compact polyhedra of dimension less than or equal to n, then K $(n+1)$-deforms to L.*

In contrast, M.M. Cohen showed that Zeeman's conjecture is false for dimensions greater than 2 [24].

To finish the section we will show that it is possible to restate the Andrews–Curtis conjecture in the context of finite spaces using Theorem 4.2.11.

Definition 11.1.3. Let X and Y be two finite T_0-spaces. We say that X n-*deforms* to Y if the later can be obtained from X by performing expansions and collapses in such a way that all the spaces involved are of height at most n.

Conjecture 11.1.4. *Let X be a homotopically trivial finite T_0-space of height 2. Then X 3-deforms to a point.*

Theorem 11.1.5. *Conjecture 11.1.4 is equivalent to the Andrews–Curtis conjecture.*

Proof. Assume Conjecture 11.1.4 is true and let K be a contractible 2-complex. Then $\mathcal{X}(K)$ is a homotopically trivial finite T_0-space of height 2 and therefore it 3-deforms to a point. By Theorem 4.2.11, K' 3-deforms to a point and then by Proposition 4.1.4, K 3-deforms to a point.

The converse follows similarly. We only have to show that X' 3-deforms to X for a finite T_0-space X of height 2. By the proof of Proposition 4.2.9, it suffices to observe that the non-Hausdorff mapping cylinder $B(h)$ of the map $h : X' \to X$ which maps a chain to its maximum, has height at most 3. $\square$

11.2   Quasi Constructible Complexes

The content of this section is in part motivated by the following example studied in Chap. 7.

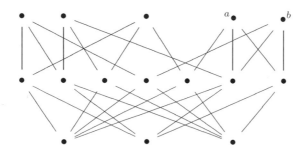

This space is the face poset of an h-regular structure of the Dunce Hat and it has no weak points. However, there are two maximal points a, b such that $U_a \cup U_b$ is contractible, and therefore $X \searrow^e Y = X \cup \{c\}$ where $a < c > b$. Now, $Y \searrow Y \smallsetminus \{a, b\}$. Thus $\mathcal{K}(X)$ 3-deforms to $\mathcal{K}(Y \smallsetminus \{a, b\})$ which has one point less than X.

In order to study the Andrews–Curtis conjecture we will be interested in complexes K whose associated finite space $\mathcal{X}(K)$ has two maximal elements a, b such that $U_a \cup U_b$ is contractible. Moreover, we will consider those complexes K such that, starting from $\mathcal{X}(K)$, one can perform repeatedly the move described above, to obtain a space with maximum, and therefore collapsible.

Let X be a finite T_0-space of height at most 2 and let a, b be two maximal elements of X such that $U_a \cup U_b$ is contractible. Then we say that there is a *qc-reduction* from X to $Y \smallsetminus \{a, b\}$ where $Y = X \cup \{c\}$ with $a < c > b$. We say that X is *qc-reducible* if we can obtain a space with a maximum by performing qc-reductions starting from X.

Note that a, b and c are all weak points of Y. Since spaces with maximum are collapsible, qc-reducible finite spaces are simple homotopy equivalent to a point. Furthermore, if X is qc-reducible, all the spaces involved in the deformation $X \searrow *$ are of height less than or equal to 3. Therefore we have the following

Remark 11.2.1. If X is qc-reducible, it 3-deforms to a point.

Example 11.2.2. The following space is collapsible but not qc-reducible. In fact we cannot perform any qc-reduction starting from X.

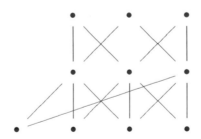

Proposition 11.2.3. *Let X be a finite T_0-space of height at most 2 and such that $H_2(X) = 0$. Let a, b be two maximal elements of X. Then the following are equivalent:*

1. *$U_a \cup U_b$ is contractible.*
2. *$U_a \cap U_b$ is nonempty and connected.*
3. *$U_a \cap U_b$ is contractible.*

Proof. The non-Hausdorff suspension $\mathbb{S}(U_a \cap U_b) = (U_a \cap U_b) \cup \{a, b\}$ is a strong deformation retract of $U_a \cup U_b$. A retraction is given by $r : U_a \cup U_b \to \mathbb{S}(U_a \cap U_b)$

with $r(x) = a$ for every $x \in U_a \smallsetminus U_b$ and $r(x) = b$ for $x \in U_b \smallsetminus U_a$. Therefore, by Proposition 2.7.3, $U_a \cup U_b$ is contractible if and only if $U_a \cap U_b$ is contractible.

Since $\mathcal{K}(X)$ has dimension at most 2, $H_3(\mathcal{K}(X), \mathcal{K}(\mathbb{S}(U_a \cap U_b))) = 0$. By the long exact sequence of homology, $H_2(\mathcal{K}(\mathbb{S}(U_a \cap U_b))) = 0$ and then $H_1(U_a \cap U_b) = 0$. Thus, if $U_a \cap U_b$ is nonempty and connected, it is contractible since $ht(U_a \cap U_b) \leq 1$. $\square$

Remark 11.2.4. If X is a contractible finite T_0-space of height at most 2, it can be proved by induction in $\#X$ that there exist two maximal elements a, b such that $U_a \cup U_b$ is contractible. However when a qc-reduction is performed, the resulting space might be not contractible.

Definition 11.2.5. A finite simplicial complex K of dimension at most 2 is said to be *quasi constructible* if K is a simplex or, recursively, if it can be written as $K = K_1 \cup K_2$ in such a way that

- K_1 and K_2 are quasi constructible,
- $K_1 \cap K_2$ is nonempty and connected, and
- No maximal simplex of K_1 is in K_2 and no maximal simplex of K_2 is in K_1.

The name of these complexes is suggested by the particular case of *constructible* complexes studied in [33].

Definition 11.2.6. A homogeneous finite simplicial complex K of dimension n is *n-constructible* if $n = 0$, if K is a simplex or if $K = K_1 \cup K_2$ where K_1 and K_2 are n-constructible and $K_1 \cap K_2$ is $(n-1)$-constructible.

A homogeneous 1-complex is 1-constructible if and only if it is connected. Therefore, 2-constructible complexes are quasi constructible. A wedge of two 2-simplices is quasi constructible but not 2-constructible. This example also shows that collapsible 2-complexes need not be 2-constructible. However we will prove below that collapsible 2-complexes are quasi constructible.

Lemma 11.2.7. *Let K be a finite simplicial complex and let K_1, K_2 be two subcomplexes such that $K_1 \cap K_2$ is a vertex v (i.e. $K = K_1 \underset{v}{\bigvee} K_2$). Then K is collapsible if and only if K_1 and K_2 are collapsible.*

Proof. Suppose $K_1 \neq v \neq K_2$. If K is collapsible and $\sigma \subseteq \sigma'$ is a collapsible pair of K such that $K \smallsetminus \{\sigma, \sigma'\}$ is collapsible, then $\sigma \subsetneq \sigma'$ is a collapsible pair of K_1 or K_2. Without loss of generality assume the first holds. Then $(K_1 \smallsetminus \{\sigma, \sigma'\}) \underset{v}{\bigvee} K_2 = K \smallsetminus \{\sigma, \sigma'\}$ is collapsible. By induction $K_1 \smallsetminus \{\sigma, \sigma'\}$ and K_2 are collapsible.

If K_1 and K_2 are collapsible, they collapse to any of their vertices. In particular $K_1 \searrow v$ and $K_2 \searrow v$. The collapses of K_1 and K_2 together show that $K \searrow v$. $\square$

Theorem 11.2.8. *Let K be a finite simplicial complex of dimension less than or equal to 2. If K is collapsible, then it is quasi constructible.*

Proof. If K is collapsible and not a point, there exists a collapsible pair $\sigma \subsetneq a\sigma$ such that $L = K \smallsetminus \{\sigma, a\sigma\}$ is collapsible. By induction L is quasi constructible. $K = L \cup a\sigma$ and $L \cap a\sigma = a\dot\sigma$ is connected. If no maximal simplex of L is a face of $a\sigma$, K is quasi constructible as we want to prove. However this might not be the case.

If $a\sigma$ is a 1-simplex and a is a maximal simplex of L, $L = a$ and then K is a 1-simplex which is quasi constructible.

Assume $a\sigma$ is a 2-simplex and let b, c be the vertices of σ.

Consider this first case: ab is a maximal simplex of L but ac is not (see Fig. 11.1). We claim that $L \smallsetminus \{ab\}$ has two connected components. Certainly, since L is contractible, from the Mayer-Vietoris sequence,

$$\widetilde{H}_1(L) \to \widetilde{H}_0(a \cup b) \to \widetilde{H}_0(ab) \oplus \widetilde{H}_0(L \smallsetminus \{ab\}) \to \widetilde{H}_0(L)$$

we deduce that $\widetilde{H}_0(L \smallsetminus \{ab\}) = \mathbb{Z}$. Therefore, there exist subcomplexes $L_1 \ni a$ and $L_2 \ni b$ of L such that $L = L_1 \underset{a}{\bigvee} ab \underset{b}{\bigvee} L_2$.

By Lemma 11.2.7, L_1 and L_2 are collapsible and therefore quasi constructible.

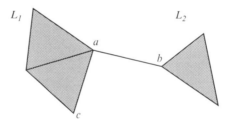

Fig. 11.1 L

Now, L_1 and $a\sigma$ are quasi constructible, $L_1 \cap a\sigma = ac$ is connected and $\{ac\}$ is not maximal in L_1 nor in $a\sigma$. Thus $L_1 \cup a\sigma$ is quasi constructible. If L_2 is just the point b, $K = L_1 \cup a\sigma$ is quasi constructible. If L_2 is not a point, $\{b\}$ is not a maximal simplex of L_2 and then $K = (L_1 \cup a\sigma) \cup L_2$ is quasi constructible since $(L_1 \cup a\sigma) \cap L_2 = b$ is connected.

The second case: ac is maximal in L but ab is not, is analogous to the first.

The third case is: ab and ac are maximal simplices of L. As above $L \smallsetminus \{ab\}$ and $L \smallsetminus \{ac\}$ have two connected components. Therefore, there exist subcomplexes L_1, L_2 and L_3 of L such that $a \in L_1$, $b \in L_2$, $c \in L_3$ and $L = L_2 \underset{b}{\bigvee} ab \underset{a}{\bigvee} L_1 \underset{a}{\bigvee} ac \underset{c}{\bigvee} L_3$. Since L is collapsible, by Lemma 11.2.7, L_i are also collapsible and by induction, quasi constructible. If $L_1 \neq a$, $L_2 \neq b$ and

$L_2 \neq c$, we prove that K is quasi constructible as follows: $a\sigma \cup L_1$ is quasi constructible since $a\sigma \cap L_1 = a$ is connected and $\{a\}$ is not maximal in $a\sigma$ nor in L_1. Then $(a\sigma \cup L_1) \cup L_2$ is quasi constructible since $(a\sigma \cup L_1) \cap L_2 = b$ is connected and $\{b\}$ is maximal in none of them. Similarly, $K = (a\sigma \cup L_1 \cup L_2) \cup L_3$ is quasi constructible. If some of the complexes L_i are just single points, this simplifies the proof since we can remove those from the writing of $K = a\sigma \cup L_1 \cup L_2 \cup L_3$.   □

On the other hand, contractible 2-constructible complexes need not be collapsible as the next example shows.

Example 11.2.9. The following example of a contractible 2-constructible and non-collapsible complex is a slight modification of one defined by Hachimori (see [33], Sect. 5.4). Let K be the 2-homogeneous simplicial complex of Fig. 11.2.

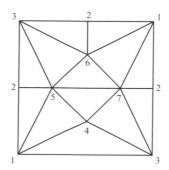

Fig. 11.2 K

This complex is 2-constructible (in fact it is shellable (see below)). For instance, one can construct it adjoining 2-simplices in the following order: $567, 457, 347, 237, 127, 167,\ 126, 236, 356, 235, 125, 145, 134$. In each adjunction both the complex and the 2-simplex are 2-constructible and their intersection is 1-constructible. Moreover, K is collapsible.

Now take two copies K_1 and K_2 of K and identify the 1-simplex 13 of both copies. The resulting complex L is contractible since K_1 and K_2 are contractible. Moreover, K_1 and K_2 are 2-constructible and their intersection is 1-constructible, therefore L is 2-constructible. On the other hand, L is not collapsible since it does not have free faces.

We will see in Corollary 11.2.11 that quasi constructible complexes 3-deform to a point. In particular this is true for this complex.

The notion of constructibility is in turn a generalization of the concept of *shellability* (see [16]), and shellable complexes are collapsible. We will not work explicitly with *shellings* in these notes. For more details on shellable complexes we refer the reader to [16] (see also [44] for an alternative definition of shellability in the context of manifolds).

Theorem 11.2.10. *Let K be a finite simplicial complex of dimension less than or equal to 2. Then the following are equivalent:*

1. *K is quasi constructible and $H_2(|K|) = 0$,*
2. *$\mathcal{X}(K)$ is qc-reducible,*
3. *K is quasi constructible and contractible.*

Proof. Let K be quasi constructible and suppose $H_2(|K|) = 0$. If K is a simplex, $\mathcal{X}(K)$ has maximum and it is qc-reducible. Otherwise, $K = K_1 \cup K_2$ where K_1 and K_2 are quasi constructible and $K_1 \cap K_2$ is connected and nonempty. Moreover the maximal simplices of K_1 are not in K_2 and vice versa. Since $H_3(|K|, |K_i|) = 0$, $H_2(|K_i|) = 0$ and by an inductive argument, $\mathcal{X}(K_i)$ is qc-reducible for $i = 1, 2$. Carrying out the same qc-reductions in $\mathcal{X}(K)$ we obtain a space Y with two maximal elements a_1 and a_2 such that $U_{a_1} \cap U_{a_2} = \mathcal{X}(K_1 \cap K_2)$ which is connected and nonempty. Moreover, $H_2(Y) = H_2(\mathcal{X}(K)) = 0$ and therefore, by Proposition 11.2.3, a last qc-reduction transforms Y in a space with maximum.

Now suppose that K is such that $\mathcal{X}(K)$ is qc-reducible. Then we can make qc-reductions to obtain a space with maximum. If $\mathcal{X}(K)$ does not have maximum, in the last step, before the last qc-reduction, one has a contractible space Y with two maximal elements a_1 and a_2. Consider the simplicial complex K_1 generated by all the maximal simplices of K that were eventually replaced by a_1 when performing the qc-reductions. Define K_2 similarly. Then, $\mathcal{X}(K_1)$ and $\mathcal{X}(K_2)$ are qc-reducible and by induction K_1 and K_2 are quasi constructible. Moreover $\mathcal{X}(K_1 \cap K_2) = U_{a_1} \cap U_{a_2}$ is connected and nonempty by Proposition 11.2.3 and then so is $K_1 \cap K_2$. Hence K is quasi constructible. On the other hand, since $\mathcal{X}(K)$ is qc-reducible, it is homotopically trivial and therefore $|K|$ is contractible.   □

In fact, the equivalence between 1 and 3 can be proved easily without going through 2 (see Remark 11.2.12).

Corollary 11.2.11. *If K is quasi constructible and contractible, it 3-deforms to a point, i.e. contractible quasi constructible complexes satisfy the Geometric Andrews–Curtis conjecture.*

Proof. If K is quasi constructible and contractible, $\mathcal{X}(K)$ is qc-reducible by Theorem 11.2.10. By Remark 11.2.1, $\mathcal{X}(K)$ 3-deforms to a point. By Theorem 4.2.11, K' 3-deforms to a point and then by Proposition 4.1.4, so does K.   □

Remark 11.2.12. By the van Kampen Theorem, quasi constructible complexes are simply connected. In particular, their reduced Euler characteristic is non-negative since their dimension is less than or equal to 2.

In the next we adapt an argument of Hachimori to show that there are many contractible 2-complexes which are not quasi constructible. The results and their proofs are essentially the same as in [33]. A vertex v of a finite complex K is a *bridge* if $K \smallsetminus v$ has more connected components than K. Following Hachimori we say that a vertex v of a finite 2-simplicial complex K is *splittable* if the graph $lk(v)$ has a bridge.

Remark 11.2.13. Suppose $K = K_1 \cup K_2$ is a 2-complex such that no maximal simplex of K_1 is in K_2 and vice versa. In this case $K_1 \cap K_2$ is a graph. Assume that there exists a vertex v which is a leaf of $K_1 \cap K_2$, i.e. $lk_{K_1 \cap K_2}(v) = v'$ is a point. We prove that v is splittable in K. Since $vv' \in K_1 \cap K_2$, vv' is not maximal in either of the subcomplexes K_1 and K_2. Let $v_i \in K_i$ such that $vv'v_i \in K_i$ for $i = 1, 2$. The vertices v_1 and v_2 are connected in $lk_K(v)$ via v'. Suppose that they are also connected in $lk_K(v) \smallsetminus v'$. Then, there exists $w \in lk_K(v) \smallsetminus v'$ such that vw is a simplex of K_1 and K_2 simultaneously. This contradicts the fact that $lk_{K_1 \cap K_2}(v) = v'$. Therefore v' is a bridge of $lk_K(v)$.

Proposition 11.2.14. *Let K be a contractible finite 2-simplicial complex with no bridges and with at most one splittable point. If K is not a 2-simplex, then it is not quasi constructible.*

Proof. Suppose that K is quasi constructible. Then there exist quasi constructible subcomplexes K_1 and K_2 as in Definition 11.2.5. $K_1 \cap K_2$ is a connected graph with more than one vertex, provided that K has no bridges. By the previous remark, it has at most one leaf and therefore it is not a tree. In particular $\widetilde{\chi}(K_1 \cap K_2) < 0$. Since K is contractible, by Remark 11.2.12 we have that

$$0 = \widetilde{\chi}(K) = \widetilde{\chi}(K_1) + \widetilde{\chi}(K_2) - \widetilde{\chi}(K_1 \cap K_2) > 0,$$

which is a contradiction. □

In particular we deduce that any triangulation of the Dunce Hat is not quasi constructible, since it has just one splittable point.

Remark 11.2.15. Recall that the Andrews–Curtis conjecture is known to be true for standard spines (see [72]). It is easy to see that standard spines have no bridges nor splittable points and therefore they are not quasi constructible. Therefore our result enlarges the class of 2-complexes for which the conjecture is known to be valid.

Any triangulation of the Dunce Hat is not quasi constructible and it is easy to see that it is not a standard spine either since it has a splittable point.

11.3 The Dual Notion of Quasi Constructibility

It seems very natural to consider the dual notion of qc-reducibility in order to obtain a larger class of complexes satisfying the Andrews–Curtis conjecture. However we will see that if K is such that $\mathcal{X}(K)^{op}$ is qc-reducible, then K is collapsible. Let X be a finite T_0-space of height at most 2 with two minimal elements a, b such that $F_a \cup F_b$ is contractible. Then we say that there is a qc^{op}-*reduction* from X to $Y \smallsetminus \{a, b\}$ where $Y = X \cup \{c\}$ with $a > c < b$. We say that X is qc^{op}-*reducible* if we can obtain a space with a minimum by carrying out qc^{op}-reductions beginning from X, or, in other words, if X^{op} is qc-reducible.

If K is a finite simplicial complex and V is a subset of vertices of K, we will denote by $st(V) \subseteq |K|$ the union of the open stars of the vertices in V, i.e.

$$st(V) = (\bigcup_{v \in V} \mathring{st}(v)),$$

where $\mathring{st}(v) = |K| \smallsetminus |K \smallsetminus v| = \bigcup_{\sigma \ni v} \mathring{\sigma} \subseteq |K|$.

We introduce the dual notion of quasi constructibility which is the following.

Definition 11.3.1. Let $K = (V_K, S_K)$ be a finite simplicial complex of dimension at most 2. We say that a subset $V \subseteq V_K$ of vertices is *quasi^{op} constructible in K* if $\#V = 1$ or if, recursively, $V = V_1 \cup V_2$ with V_i quasi^{op} constructible in K for $i = 1, 2$, $V_1 \cap V_2 = \emptyset$ and $st(V_1) \cap st(V_2)$ is a connected nonempty subspace of the geometric realization $|K|$.

The complex K is said to be *quasi^{op} constructible* if V_K is quasi^{op} constructible in K.

In order to understand the topology of $st(V_1) \cap st(V_2)$, we will generalize the result that says that $\mathcal{X}(K)$ is a finite model of K, giving an alternative proof of this fact.

Theorem 11.3.2. *Let K be a finite simplicial complex and let $Y \subseteq S_K$ be a subset of simplices of K. Let $X = \bigcup_{\sigma \in Y} \mathring{\sigma} \subseteq |K|$ and let $f : X \to Y \subseteq \mathcal{X}(K)^{op}$ be the map defined by $f(x) = \sigma$ if $x \in \mathring{\sigma}$. Then, f is a weak homotopy equivalence.*

Proof. We first note that f is continuous. If $\sigma \in Y$,

$$f^{-1}(U_\sigma) = \bigcup_{\sigma \subseteq \tau \in Y} \mathring{\tau} = (\bigcup_{\sigma \subseteq \tau \in S_K} \mathring{\tau}) \cap X = X \smallsetminus |\sigma^c|$$

is open in X since σ^c is a subcomplex of K. To prove that f is a weak homotopy equivalence we use the Theorem of McCord 1.4.2. We only have to show that $f^{-1}(U_\sigma)$ is contractible. In fact, $\mathring{\sigma}$ is a strong deformation retract of $f^{-1}(U_\sigma)$. Let $x \in \mathring{\tau}$ with $\sigma \subseteq \tau \in Y$, $x = t\alpha + (1-t)\beta$ for some $0 < t \leq 1$, $\alpha \in \mathring{\sigma}$ and $\beta \in (\tau \smallsetminus \sigma)^\circ$. Define $r : f^{-1}(U_\sigma) \to \mathring{\sigma}$ by $r(x) = \alpha$. Then r is a retraction and $H : f^{-1}(U_\sigma) \times I \to f^{-1}(U_\sigma)$, given by $H(x, s) = (1-s)x + s\alpha$, defines a homotopy between $1_{f^{-1}(U_\sigma)}$ and ir. □

Proposition 11.3.3. *Let K be a finite T_0-space of height at most 2. Then K is quasi^{op} constructible and contractible if and only if $\mathcal{X}(K)$ is qc^{op}-reducible.*

Proof. Suppose $|K|$ is contractible. We prove that if $V \subseteq V_K$ is quasi^{op} constructible in K, then $\bigcup_{v \in V} F_{\{v\}} \subseteq \mathcal{X}(K)$ is qc^{op}-reducible. If $\#V = 1$, $\bigcup_{v \in V} F_{\{v\}}$ has minimum and there is nothing to do. Assume that $V = V_1 \cup V_2$ where V_1 and V_2 are disjoint and quasi^{op} constructible in K, and $st(V_1) \cap st(V_2)$ is connected and nonempty. By induction $\bigcup_{v \in V_1} F_{\{v\}}$ and $\bigcup_{v \in V_2} F_{\{v\}}$ are qc^{op}-reducible. Then $\bigcup_{v \in V} F_{\{v\}}$ qc^{op}-reduces to a space X with two minimal elements a_1 and a_2. Moreover, $F_{a_1} \cap F_{a_2} = \{\sigma \in S_K \mid$ there exist $v_1 \in V_1$ and $v_2 \in V_2$ with $v_1, v_2 \in \sigma\}$ is weak homotopy equivalent to $st(V_1) \cap st(V_2)$ by Theorem 11.3.2. In particular, $F_{a_1} \cap F_{a_2}$ is connected and nonempty, and since $\mathcal{X}(K)$ is homotopically trivial, by Proposition 11.2.3, X is contractible. Therefore a last qc^{op}-reductions transforms X into a space with minimum, so $\bigcup_{v \in V} F_{\{v\}}$ is qc^{op}-reducible. Now, if in addition K is quasi^{op}-constructible, V_K is quasi^{op} constructible in K and then $\mathcal{X}(K) = \bigcup_{v \in V_K} F_{\{v\}}$ is qc^{op}-reducible.

Conversely, let $V \in V_K$ be a subset of vertices of K. We will prove that if $\bigcup_{v \in V} F_{\{v\}} \subseteq \mathcal{X}(K)$ is qc^{op}-reducible, then V is quasi^{op} constructible in K. If $\#V = 1$ there is nothing to prove. In other case, before the last step we will have reduced $\bigcup_{v \in V} F_{\{v\}}$ into a contractible space X with two minimal points a_1 and a_2. Let V_i be the subset of V of vertices that were eventually replaced by a_i for $i = 1, 2$. Then $\bigcup_{v \in V_i} F_{\{v\}}$ is qc^{op}-reducible and by induction V_i is quasi^{op} constructible for $i = 1, 2$. By Theorem 11.3.2, $st(V_1) \cap st(V_2)$ is weak homotopy equivalent to $F_{a_1} \cap F_{a_2}$ which is connected and nonempty by Proposition 11.2.3. Then V is quasi^{op} constructible in K.

Finally, applying this result to $V = V_K$ we deduce that if $\mathcal{X}(K)$ is qc^{op}-reducible, then K is quasi^{op} constructible. In this case $\mathcal{X}(K)$ is homotopically trivial and then $|K|$ is contractible. $\qquad\square$

In particular, we deduce that if K is quasi^{op} constructible and contractible, it 3-deforms to a point. Unfortunately, this does not enlarge the class of complexes satisfying the Andrews–Curtis conjecture, since quasi^{op} constructible complexes are collapsible as we will see.

Lemma 11.3.4. *Let K be a finite simplicial complex of dimension less than or equal to 2. If $V \subseteq V_K$ is quasi^{op} constructible in K, then $\widetilde{\chi}(st(V)) \geq 0$.*

Proof. If $\#V = 1$, $st(V)$ is contractible and then $\widetilde{\chi}(st(V)) = 0$. Suppose that $V = V_1 \cup V_2$ where V_1 and V_2 are disjoint, quasi^{op} constructible in K and such that $st(V_1) \cap st(V_2)$ is connected and nonempty. By induction,

$$\widetilde{\chi}(st(V)) = \widetilde{\chi}(st(V_1)) + \widetilde{\chi}(st(V_2)) - \widetilde{\chi}(st(V_1) \cap st(V_2)) \geq -\widetilde{\chi}(st(V_1) \cap st(V_2)).$$

By Theorem 11.3.2, $st(V_1) \cap st(V_2)$ is weak homotopy equivalent to $\overline{V}_1 \cap \overline{V}_2 \subseteq \mathcal{X}(K)$ which is a finite T_0-space of height at most 1. Since it is connected and nonempty, $\widetilde{\chi}(st(V_1) \cap st(V_2)) = \widetilde{\chi}(\overline{V}_1 \cap \overline{V}_2) \leq 0$ and then $\widetilde{\chi}(st(V)) \geq 0$.
$\square$

Theorem 11.3.5. *Let K be a contractible quasi^{op} constructible simplicial complex. Then K is collapsible.*

Proof. If $K = *$, there is nothing to do. Suppose $V_K = V_1 \cup V_2$ with $V_1 \cap V_2 = \emptyset$, V_1 and V_2 quasi^{op} constructible in K and $st(V_1) \cap st(V_2)$ nonempty and connected. Since $|K|$ is contractible,

$$0 = \widetilde{\chi}(|K|) = \widetilde{\chi}(st(V_1)) + \widetilde{\chi}(st(V_2)) - \widetilde{\chi}(st(V_1) \cap st(V_2)).$$

By Lemma 11.3.4, $\widetilde{\chi}(st(V_i)) \geq 0$ for $i = 1, 2$ and then $\widetilde{\chi}(\overline{V}_1 \cap \overline{V}_2) = \widetilde{\chi}(st(V_1) \cap st(V_2)) \geq 0$. Moreover, $\overline{V}_1 \cap \overline{V}_2 \subseteq \mathcal{X}(K)$ is nonempty, connected and its height is less than or equal to 1. Therefore, it is contractible. In particular, there exists a simplex $\sigma \in K$ which is a leaf (maybe the unique vertex) of the graph $\mathcal{K}(\overline{V}_1 \cap \overline{V}_2)$. We claim that σ is not a 2-simplex, because if that was the case, it would have two of its vertices a, b in V_i and the third c in V_j for $i \neq j$. Then $\{a, c\}$ and $\{b, c\}$ would be covered by σ in $\overline{V}_1 \cap \overline{V}_2$ contradicting the fact that σ is a leaf of $\mathcal{K}(\overline{V}_1 \cap \overline{V}_2)$. Thus σ is a 1-simplex.

Let $a \in V_1$ and $b \in V_2$ be the vertices of σ. Since σ is a leaf of $\mathcal{K}(\overline{V}_1 \cap \overline{V}_2)$, we consider two different cases:

(1) $\overline{V}_1 \cap \overline{V}_2 = \{\sigma\}$ or
(2) $\sigma \in K$ is a free face of a simplex $\sigma' = \{a, b, c\} \in K$.

We study first the case (1). For $i = 1, 2$, let K_i be the full subcomplex of K spanned by the vertices of V_i. Then $K = K_1 \cup K_2 \cup \{\sigma\} = K_1 \underset{a}{\bigvee} \sigma \underset{b}{\bigvee} K_2$. Since K is contractible, K_1 and K_2 are contractible as well. Moreover, since V_i is quasi^{op} constructible in K, it is also quasi^{op} constructible in K_i. Note that if V and V' are subsets of V_i, then $st_{K_i}(V) \cap st_{K_i}(V') = st_K(V) \cap st_K(V')$. Thus, K_1 and K_2 are contractible and quasi^{op} constructible. By induction, they are collapsible. Therefore $K = K_1 \underset{a}{\bigvee} \sigma \underset{b}{\bigvee} K_2$ is also collapsible.

Now we consider the second case (2). Let $L = K \smallsetminus \{\sigma, \sigma'\}$. By hypothesis $K \searrow L$. We claim that L is quasi^{op} constructible. To prove that, we will show first that V_1 and V_2 are quasi^{op} constructible in L. We prove by induction that if $V \subseteq V_1$ is quasi^{op} constructible in K, then it also is in L. If $\#V = 1$ this is trivial. Suppose $V = V' \cup V''$ with V' and V'' disjoint, quasi^{op} constructible in K and such that $st_K(V') \cap st_K(V'') \overset{we}{\approx} \overline{V'}^{\mathcal{X}(K)} \cap \overline{V''}^{\mathcal{X}(K)}$ is nonempty and connected. By induction V' and V'' are quasi^{op} constructible in L. We have to show that $\overline{V'}^{\mathcal{X}(L)} \cap \overline{V''}^{\mathcal{X}(L)} = (\overline{V'}^{\mathcal{X}(K)} \cap \overline{V''}^{\mathcal{X}(K)}) \smallsetminus \{\sigma, \sigma'\}$ is nonempty and connected.

Since σ has only one vertex in V_1, it cannot have a vertex in V' and other in V''. Therefore, $\sigma \notin \overline{V'}^{\mathcal{X}(K)} \cap \overline{V''}^{\mathcal{X}(K)}$. If $\sigma' \notin \overline{V'}^{\mathcal{X}(K)} \cap \overline{V''}^{\mathcal{X}(K)}$, then $\overline{V'}^{\mathcal{X}(L)} \cap \overline{V''}^{\mathcal{X}(L)} = (\overline{V'}^{\mathcal{X}(K)} \cap \overline{V''}^{\mathcal{X}(K)})$ is nonempty and connected. If $\sigma' \in \overline{V'}^{\mathcal{X}(K)} \cap \overline{V''}^{\mathcal{X}(K)}$, then $c \in V_1$ and σ' covers just one element of $\overline{V'}^{\mathcal{X}(K)} \cap \overline{V''}^{\mathcal{X}(K)}$, which is $\{a, c\}$. Hence, σ' is a down beat point of $\overline{V'}^{\mathcal{X}(K)} \cap \overline{V''}^{\mathcal{X}(K)}$ and in particular $\overline{V'}^{\mathcal{X}(L)} \cap \overline{V''}^{\mathcal{X}(L)}$ is homotopy equivalent to $\overline{V'}^{\mathcal{X}(K)} \cap \overline{V''}^{\mathcal{X}(K)}$. Then, it is nonempty and connected and therefore V is quasi^{op} constructible in L.

Since V_1 is quasi^{op} constructible in K it follows that it is quasi^{op} constructible in L. Analogously, V_2 is quasi^{op} constructible in L.

Now, by assumption $st_K(V_1) \cap st_K(V_2) \overset{we}{\approx} \overline{V_1}^{\mathcal{X}(K)} \cap \overline{V_2}^{\mathcal{X}(K)}$ is nonempty and connected. Since σ is a free face of K, it is an up beat point of $\overline{V_1}^{\mathcal{X}(K)} \cap \overline{V_2}^{\mathcal{X}(K)}$. On the other hand, σ' is a down beat point of $\overline{V_1}^{\mathcal{X}(K)} \cap \overline{V_2}^{\mathcal{X}(K)} \smallsetminus \{\sigma\}$ since there is a 1-face of σ' with both vertices in V_1 or in V_2. Hence, $\overline{V_1}^{\mathcal{X}(L)} \cap \overline{V_2}^{\mathcal{X}(L)} = \overline{V_1}^{\mathcal{X}(K)} \cap \overline{V_2}^{\mathcal{X}(K)} \smallsetminus \{\sigma, \sigma'\}$ is a strong deformation retract of $\overline{V_1}^{\mathcal{X}(K)} \cap \overline{V_2}^{\mathcal{X}(K)}$, and then it is connected and nonempty. Thus, $V_L = V_1 \cup V_2$ is quasi^{op} constructible in L, or in other words, L is quasi^{op} constructible.

Since $K \searrow^e L$, L is contractible and quasi^{op} constructible. By induction L is collapsible and therefore, so is K. $\qquad\square$

The converse of this result is false as we prove in the next example.

Example 11.3.6. The complex K studied in Example 11.2.9 is a collapsible homogeneous 2-complex with a unique free face. We prove that a complex satisfying these hypotheses cannot be quasi^{op} constructible.

Suppose that K is quasi^{op} constructible. Since K has more than one vertex, V_K can be written as a disjoint union of quasi^{op} constructible subsets V_1 and V_2 in K such that $\overline{V}_1 \cup \overline{V}_2$ is contractible. The case (1) of the proof of Theorem 11.3.5 cannot occur since K is homogeneous. Therefore, $\mathcal{K}(\overline{V}_1 \cap \overline{V}_2)$ has dimension exactly 1 and it is a tree. Then, it has at least two leaves, which must be 1-simplices and free faces of K. However this is absurd since K has only one free face.

The qc and qc^{op}-reductions studied in this chapter conclude the list of reduction methods introduced in this work. The last example of this book is a homotopically trivial finite space X which cannot be reduced via any of these methods. Its aim is to motivate the development of new methods of reduction. Although this example is not directly related to the Andrews–Curtis conjecture, it is connected to the methods studied in this chapter and to several ideas developed throughout these notes.

Example 11.3.7. Consider the following pentagon whose edges are identified as indicated by the arrows.

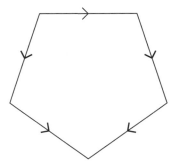

This CW-complex is contractible since the attaching map of the 2-cell is a homotopy equivalence $S^1 \to S^1$. We endow this space with an h-regular structure K as follows

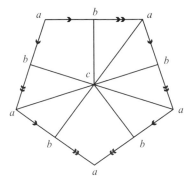

Since K is contractible, $\mathcal{X}(K)$ is a homotopically trivial finite space of 21 points by Theorem 7.1.7. It is easy to check that $\mathcal{X}(K)$ has no weak points, and therefore it does not have γ-points either since its height is 2. In fact no h-regular CW-complex has down weak points and it is not hard to see that the 0-cells of this example are not up weak points. We only have to show that $\hat{F}_a$, $\hat{F}_b$ and $\hat{F}_c$ are not contractible, but this is clear since their associated graphs contain a cycle.

It is not possible to make a qc-reduction on $\mathcal{X}(K)$, since for any 2-cells e, e' of K, $\bar{e} \cap \bar{e'} \subseteq K$ is not connected. It can also be proved that no qc^{op}-reduction can be made in $\mathcal{X}(K)$ since the subspaces $F_a \cap F_b, F_a \cap F_c, F_b \cap F_c \subseteq \mathcal{X}(K)$ are nonconnected.

Osaki's reduction methods 6.1.1 and 6.1.2 are not applicable either.

On the other hand we know that it is possible to obtain the singleton starting from $\mathcal{X}(K)$ and performing expansions and collapses.

Chapter 12
Appendix

This appendix is intended to recall some of the basic notions and properties of simplicial complexes and CW-complexes which are used but not explicitly explained in the main text. The reader not familiar with concepts such as simplicial approximations and adjunction spaces, could find this appendix useful. However, more complete expositions on these subjects can be found in Spanier's book [75, Chap. 3] and in Munkres' [61, Chaps. 1 and 2]. Standard references for CW-complexes are also [28, 38, 45].

A.1 Simplicial Complexes

A *simplicial complex* K consists of a set V_K, called the set of vertices, and a set S_K of finite nonempty subsets of V_K, which is called the set of simplices, satisfying that any subset of V of cardinality one is a simplex and any nonempty subset of a simplex is a simplex. By abuse of notation we will write $v \in K$ and $\sigma \in K$ if $v \in V_K$ and $\sigma \in S_K$. Many times, as it is the custom, we will identify a simplicial complex with its set of simplices.

If a simplex σ is contained in another simplex τ, it is called a *face* of τ, and it is a *proper face* if in addition $\sigma \neq \tau$. A simplex with $n+1$ vertices is called an *n-simplex*, and we say that its *dimension* is n. Note that the vertices of K correspond to the 0-simplices. The *dimension* of K is the supremum of the dimensions of its simplices. If K is empty, its dimension is -1 and if K contains simplices of arbitrary large dimension, its dimension is infinite. An *n-complex* is a simplicial complex of dimension n. The maximal simplices (those which are not proper faces of any other simplex) are sometimes called *facets*. A finite dimensional simplicial complex is called *homogeneous* (or *pure*) if all its maximal simplices have the same dimension. A *subcomplex* of a simplicial complex K is a simplicial complex L such that $V_L \subseteq V_K$ and $S_L \subseteq S_K$. A subcomplex $L \subseteq K$ is said to be *full* if any simplex of K with

J.A. Barmak, *Algebraic Topology of Finite Topological Spaces and Applications*, Lecture Notes in Mathematics 2032, DOI 10.1007/978-3-642-22003-6, © Springer-Verlag Berlin Heidelberg 2011

all its vertices in L is also a simplex of L. In this case we say that L is the full subcomplex of K spanned by the vertices $v \in V_L$.

Given a simplex $\sigma = \{v_0, v_1, \ldots, v_n\}$ of dimension n, the *closed simplex* $\overline{\sigma}$ is the set of formal convex combinations $\sum\limits_{i=0}^{n} \alpha_i v_i$ with $\alpha_i \geq 0$ for every $0 \leq i \leq n$ and $\sum \alpha_i = 1$. A closed simplex is a metric space with the metric d given by

$$d(\sum_{i=0}^{n} \alpha_i v_i, \sum_{i=0}^{n} \beta_i v_i) = \sqrt{\sum_{i=0}^{n} (\alpha_i - \beta_i)^2}.$$

The *geometric realization* $|K|$ of a simplicial complex K is the set of formal convex combinations $\sum\limits_{v \in K} \alpha_v v$ such that $\{v \mid \alpha_v > 0\}$ is a simplex of K. Therefore, $|K|$ can be regarded as the union of the closed simplices $\overline{\sigma}$ with $\sigma \in K$. The topology of $|K|$ is the final (coherent) topology with respect to the closed simplices. In other words, a set $U \subseteq |K|$ is open (resp. closed) if and only if $U \cap \overline{\sigma}$ is open (resp. closed) in the metric space $\overline{\sigma}$ for every $\sigma \in K$.

The *support* (or *carrier*) of a point $x = \sum\limits_{v \in K} \alpha_v v \in |K|$ is the simplex $support(x) = \{v \mid \alpha_v > 0\}$. If σ is a simplex, the *open simplex* $\overset{\circ}{\sigma}$ is the subset of $\overline{\sigma}$ of points whose support is exactly σ. Note that if two points $x, y \in |K|$ lie in the same closed simplex, then the convex combination $tx + (1-t)y$ is a well defined element in $|K|$. If $L \subseteq K$, $|L|$ is a closed subspace of $|K|$. It is not hard to prove that the topology of the set $\overline{\sigma}$ as a subspace of $|K|$ is the original metric topology on $\overline{\sigma}$. Moreover, if K is a finite simplicial complex (i.e. with a finite number of vertices), the topology of $|K|$ coincides with the metric topology defined as before

$$d(\sum_{v \in K} \alpha_v v, \sum_{v \in K} \beta_v v) = \sqrt{\sum_{v \in K} (\alpha_v - \beta_v)^2}.$$

Moreover, in this case $|K|$ can be imbedded in $\mathbb{R}^n$ for some $n \in \mathbb{N}$.

It is easy to prove that if U is an open subspace of $|K|$, then it has the final topology with respect to the subspaces $U \cap \overline{\sigma} \subseteq \overline{\sigma}$.

A *polyhedron* is the geometric realization of a simplicial complex and a *triangulation* of a polyhedron X is a simplicial complex K whose geometric realization is homeomorphic to X. Any polyhedron is a Hausdorff space.

Since $|K|$ has the final topology with respect to its closed simplices, a map f from $|K|$ to a topological space X is continuous if and only if each of the restrictions $f|_{\overline{\sigma}} : \overline{\sigma} \to X$ is continuous. Moreover, by the exponential law, it can be shown that a map $H : |K| \times I \to X$ is continuous if and only if $H|_{\overline{\sigma} \times I} : \overline{\sigma} \times I \to X$ is continuous for each $\sigma \in K$.

A *simplicial map* $\varphi : K \to L$ between two simplicial complexes K and L is a vertex map $V_K \to V_L$ that sends simplices into simplices. A simplicial map

$\varphi : K \to L$ induces a well defined continuous map $|\varphi| : |K| \to |L|$ between the geometric realizations defined by $|\varphi|(\sum_{v \in K} \alpha_v v) = \sum_{v \in K} \alpha_v \varphi(v)$.

Lemma A.1.1. *Let K be a simplicial complex and let F be a compact subset of $|K|$. Then there exists a finite subcomplex L of K such that $F \subseteq |L|$.*

Proof. Take one point in $F \cap \overset{\circ}{\sigma}$ for every open simplex intersecting F. Denote by D the set of all these points. Let $A \subseteq D$. Since the intersection of A with each closed simplex is finite, it is closed, and then A is closed in $|K|$. Therefore D is discrete and compact, and, in particular, finite. Thus, F intersects only finitely many open simplices. The complex L generated by (i.e. the smallest complex containing) the simplices σ such that $\overset{\circ}{\sigma}$ intersects F is a finite subcomplex of K which satisfies the required property.   □

Proposition A.1.2. *Let K and L be two simplicial complexes and let $f, g : |K| \to |L|$ be two continuous maps such that for every $x \in |K|$ there exists $\sigma \in L$ with $f(x), g(x) \in \overline{\sigma}$. Then f and g are homotopic.*

Proof. The map $H : |K| \times I \to |L|$ given by $H(x,t) = tg(x) + (1-t)f(x)$ is well defined because $g(x)$ and $f(x)$ lie in a same closed simplex. In order to prove that H is continuous it suffices to show that it is continuous in $\overline{\sigma} \times I$ for every $\sigma \in K$. If $\sigma \in K$, $\overline{\sigma}$ is compact and therefore $f(\overline{\sigma})$ and $g(\overline{\sigma})$ are compact. By Lemma A.1.1, $f(\overline{\sigma})$ is contained in the geometric realization of a finite subcomplex L_1 and $g(\overline{\sigma}) \subseteq |L_2|$ for a finite subcomplex $L_2 \subseteq L$. Therefore, $H(\overline{\sigma} \times I)$ is contained in the realization of a finite subcomplex M of L, namely the full subcomplex spanned by the vertices of L_1 and L_2. We have to show then that $H|_{\overline{\sigma} \times I} : \overline{\sigma} \times I \to |M|$ is continuous, where M is a finite simplicial complex. But this is clear since both $\overline{\sigma}$ and $|M|$ have the metric topology.

$$d(H(x,t), H(y,s)) \leq d(tg(x) + (1-t)f(x), sg(x) + (1-s)f(x))$$

$$+d(sg(x) + (1-s)f(x), sg(y) + (1-s)f(y))$$

$$\leq 2|t - s| + d(f(x), f(y)) + d(g(x), g(y)).$$

Therefore, the continuity of H follows from that of f and g.   □

The homotopy H used in the proof of Proposition A.1.2 is called the *linear homotopy* from f to g.

Two simplicial maps $\varphi, \psi : K \to L$ are said to be *contiguous* if for every $\sigma \in K$, $\varphi(\sigma) \cup \psi(\sigma)$ is a simplex of L. In this case, $|\varphi|$ and $|\psi|$ satisfy the hypothesis of Proposition A.1.2, since if $x \in \overline{\sigma}$, both $|\varphi|(x)$ and $|\psi|(x)$ lie in $\varphi(\sigma) \cup \psi(\sigma)$. Therefore we deduce the following

Corollary A.1.3. *If φ and ψ are two contiguous maps, $|\varphi|$ and $|\psi|$ are homotopic.*

A *simplicial cone with apex v* is a simplicial complex K with a vertex v satisfying that for every simplex σ of K, $\sigma \cup \{v\}$ is also a simplex.

Corollary A.1.4. *If K is a simplicial cone, $|K|$ is contractible.*

Proof. Let v be the (an) apex of K. The simplicial map that sends every vertex to v is contiguous to the identity by definition of cone. Therefore, by Corollary A.1.3, the identity of $|K|$ is homotopic to a constant. ☐

Given a simplicial complex K, its *barycentric subdivision* K' is the following simplicial complex. The vertices of K' are the simplices of K, and a simplex of K' is a chain of simplices of K, i.e. a set $\{\sigma_0, \sigma_1, \ldots, \sigma_n\}$ of simplices of K such that $\sigma_0 \subsetneqq \sigma_1 \subsetneqq \ldots \subsetneqq \sigma_n$. The *barycenter* of a simplex $\sigma \in K$ is the point $b(\sigma) = \sum_{v \in \sigma} \frac{v}{\#\sigma} \in |K|$. The *linear map* $s_K : |K'| \to |K|$ defined by $s_K(\sigma) = b(\sigma)$ is a homeomorphism. By linear we mean a map that preserves convex combinations. The spaces $|K'|$ and $|K|$ are usually identified by means of the map s_K in such a way that s_K becomes the identity map.

A simplicial map $\varphi : K \to L$ is said to be a *simplicial approximation* of a continuous map $f : |K| \to |L|$ if $f(x) \in \overline{\sigma}$ implies $|\varphi|(x) \in \overline{\sigma}$ for every $x \in |K|$. Note that in this situation f and $|\varphi|$ are homotopic by Proposition A.1.2.

Proposition A.1.5. *A vertex map $\varphi : K' \to K$ is a simplicial approximation to the identity $s_K : |K'| \to |K|$ if and only if $\varphi(\sigma) \in \sigma$ for every $\sigma \in K$.*

Proof. Suppose that φ is a simplicial approximation to the identity. If σ is a vertex of K', $s_K(\sigma) = b(\sigma) \in \overline{\sigma}$, then $|\varphi|(\sigma)$ must be contained in $\overline{\sigma}$ as well. Therefore $\varphi(\sigma)$ is a vertex of σ.

Conversely, suppose $\varphi : K' \to K$ is a vertex map in the hypothesis of the proposition. If $\sigma_0 \subsetneqq \sigma_1 \subsetneqq \ldots \subsetneqq \sigma_n$ is a chain of simplices of K, then $\varphi(\{\sigma_0, \sigma_1, \ldots, \sigma_n\}) \subseteq \sigma_n$. Therefore φ is a simplicial map. Moreover if

$$x = \sum_{i=0}^{n} \alpha_i \sigma_i,$$

with $\alpha_i > 0$ for every i, then

$$s_K(x) = \sum_{i=0}^{n} \alpha_i \sum_{v \in \sigma_i} \frac{v}{\#\sigma_i} \in \overset{\circ}{\sigma}_n.$$

On the other hand, $|\varphi|(x) = \sum_{i=0}^{n} \alpha_i \varphi(\sigma_i) \in \overline{\sigma}_n$. Thus, φ is a simplicial approximation of s_K. ☐

As an immediate consequence we deduce that there exist simplicial approximations to the identity.

The *n-th barycentric subdivision* of K is defined recursively $K^{(n)} = (K^{(n-1)})'$. A simplicial approximation to the identity $1_{|K|} : |K^{(n)}| \to |K|$ is in this case a simplicial approximation of the map $s_K s_{K'} \ldots s_{K^{(n-1)}} : |K^{(n)}| \to |K|$. If $f : |K| \to |L|$ is a continuous map, a simplicial map $\varphi : K^{(n)} \to L$ is called an approximation of f if it is a simplicial approximation of $f s_K s_{K'} \ldots s_{K^{(n-1)}}$.

The proof of the following result on simplicial approximations can be found in [75, Corollary 3.4.5, Lemma 3.5.4].

Proposition A.1.6.

1. *The composition of simplicial approximations of two maps is a simplicial approximation of the composition of those maps.*
2. *Two simplicial approximations to the same map are contiguous.*

Two simplicial maps $\varphi, \psi : K \to L$ are said to be in the same *contiguity class* if there is a sequence of simplicial maps $\varphi = \varphi_0, \varphi_1, \ldots, \varphi_k = \psi$ from K to L, such that φ_i and φ_{i+1} are contiguous for every $0 \leq i < k$.

The following result is known as the Simplicial Approximation Theorem. Its proof can be found in [75, Theorems 3.4.8 and 3.5.6].

Theorem A.1.7. *Let K be a finite simplicial complex and L a simplicial complex. Given any continuous map $f : |K| \to |L|$ there exist $n \in \mathbb{N}$ and a simplicial approximation $\varphi : K^{(n)} \to L$ to f. Moreover, if $f, g : |K| \to |L|$ are homotopic, there exist $n \in \mathbb{N}$ and simplicial approximations $\varphi, \psi : K^{(n)} \to L$ to f and g in the same contiguity class.*

A.2 CW-Complexes and a Gluing Theorem

If X, Y and Z are three topological spaces, and $f : X \to Y$, $g : X \to Z$ are continuous maps, the *pushout* of the diagram

$$
\begin{array}{ccc}
X & \xrightarrow{\ f\ } & Y \\
\downarrow{\scriptstyle g} & & \\
Z & &
\end{array}
$$

is a space P together with maps $\overline{f} : Z \to P$ and $\overline{g} : Y \to P$ such that $\overline{f}g = \overline{g}f$ and with the following universal property: for any space Q and maps $\widetilde{f} : Z \to Q$ and $\widetilde{g} : Y \to Q$ such that $\widetilde{f}g = \widetilde{g}f$, there exists a unique map $h : P \to Q$ such that $h\overline{f} = \widetilde{f}$ and $h\overline{g} = \widetilde{g}$.

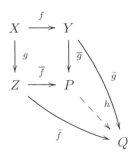

It is not hard to see that the space P is unique up to homeomorphism and in fact it can be characterized as the space $P = (Z \sqcup Y)/ \sim$, where $\sim$ is the relation that identifies $f(x)$ with $g(x)$ for every $x \in X$. The maps $\overline{f}$ and $\overline{g}$ are the canonical inclusions into the disjoint union $Z \sqcup Y$ composed with the quotient map.

For example, if A is a subspace of a space X, the quotient X/A is the pushout of the diagram

If A and B are two closed (or two open) subspaces of a space X and $X = A \cup B$, then X is the pushout of $A \hookleftarrow A \cap B \hookrightarrow B$.

A *topological pair* is an ordered pair of spaces (X, A) with A a subspace of X. In the next definition the inclusions $A \hookrightarrow A \times I$ and $X \hookrightarrow X \times I$ of the spaces A and X in the bases of their cylinders will be denoted by i_0 and j_0 respectively.

Definition A.2.1. A topological pair (X, A) is said to have the *homotopy extension property* if for any space Y and maps $H : A \times I \to Y$, $f : X \to Y$ such that $Hi_0 = f|_A$, there exists a map $\overline{H} : X \times I \to Y$ such that $\overline{H}j_0 = f$ and $\overline{H}|_{A \times I} = H$.

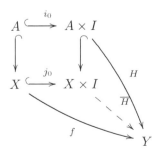

If A is a subspace of a space X, the inclusion $A \hookrightarrow X$ is said to be a *closed cofibration* if A is closed in X and (X, A) has the homotopy extension property.

Definition A.2.2. If $A \subseteq X$, and the inclusion $A \hookrightarrow X$ is a closed cofibration, the pushout Z of a diagram

$$
\begin{array}{ccc}
A & \xrightarrow{\ f\ } & Y \\
\cap \downarrow & & \\
X & &
\end{array}
$$

is called the *adjunction space* of X to Y by f. In this case, it can be proved that Y is a closed subspace of Z (see [21]).

When $X = D^n$ is a disk and $A = S^{n-1}$ is its boundary, we say that Z is constructed from Y by *adjoining an n-cell*. More generally, if $X = \bigsqcup\limits_{\alpha \in \Lambda} D^n$ is a union of n-dimensional disks indexed by an arbitrary set Λ and $A = \bigsqcup\limits_{\alpha \in \Lambda} S^{n-1}$, we say that Z is obtained from Y by adjoining n-cells.

Definition A.2.3. A *CW-structure* for a topological space X is a filtration of X by subspaces $X^0 \subseteq X^1 \subseteq \ldots$, where X^0 is a discrete space, X^n is constructed from X^{n-1} by adjoining n-cells and X is the union of the spaces X^n, $n \geq 0$, with the final (coherent) topology. The subspace X^n is called the *n-skeleton* of X.

A *CW-complex* is a space X endowed with a CW-structure. Note that, since X^n is obtained from X^{n-1} by adjoining n-cells, there is a pushout

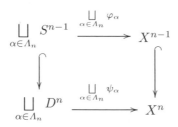

The image of the map $\psi_\alpha : D^n \to X$ is called the *closed cell* $\overline{e^n_\alpha}$. The image of $\varphi_\alpha : S^{n-1} \to X$ is the *boundary* $\dot{e}^n_\alpha$ of the cell and the (*open*) *cell* e^n_α is concretely the subspace $e^n_\alpha = \overline{e^n_\alpha} \setminus \dot{e}^n_\alpha$, which is homeomorphic to the interior of the disk D^n. The maps φ_α and ψ_α are called the *attaching map* and the *characteristic map* of the cell e^n_α, respectively. For us, the attaching and characteristic maps will be part of the structure of the CW-complex. For some authors, though, the CW-structure consists only of the filtration by skeleta. In that case the characteristic maps are not part of the structure, only their existence is required.

A cell e_α^n is called an *n-cell* or cell of *dimension n*. The *dimension* of a CW-complex X is -1 if it is empty, n if $X^n \neq X^{n-1}$ and $X^m = X^n$ for every $m \geq n$, and infinite if $X^n \neq X$ for every n.

Simplicial complexes are CW-complexes. Their cells are the open simplices. Many properties of polyhedra hold in fact for CW-complexes. For instance, any CW-complex has the final topology with respect to its closed cells and every CW-complex is a Hausdorff space.

A *subcomplex* of a CW-complex X is a closed subspace of X which is a union of cells of X. The following is a basic result about CW-complexes. A proof can be found in [75, 7.6.12] or [28, Corollary 1.4.7].

Theorem A.2.4. *If A is a subcomplex of a CW-complex X, the inclusion $A \hookrightarrow X$ is a closed cofibration.*

The following gluing theorem appears for instance in [21, 7.5.7, Corollary 2].

Theorem A.2.5. *Suppose that the following diagram is a pushout of topological spaces*

$$
\begin{array}{ccc}
A & \xrightarrow{\ f\ } & Y \\
\downarrow & & \downarrow \\
X & \xrightarrow{\ \overline{f}\ } & Z
\end{array}
$$

in which $A \hookrightarrow X$ is a closed cofibration and $f : A \to Y$ is a homotopy equivalence. Then $\overline{f} : X \to Z$ is a homotopy equivalence.

The following are two applications that show how to use the gluing theorem together with Theorem A.2.4 to study the homotopy type of CW-complexes and polyhedra.

Proposition A.2.6. *Let K be a simplicial complex and let v be a vertex of K. If the link $|lk(v)|$ is contractible, $|K|$ and $|K \smallsetminus v|$ are homotopy equivalent.*

Proof. Consider the following diagram

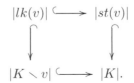

It is a pushout because $|st(v)| \cup |K \smallsetminus v| = |K|$, $|st(v)| \cap |K \smallsetminus v| = |lk(v)|$ and both $|st(v)|$ and $|K \smallsetminus v|$ are closed subspaces of $|K|$. Moreover, $|lk(v)| \hookrightarrow |K \smallsetminus v|$ is a cofibration by Theorem A.2.4 and, since $|lk(v)| \hookrightarrow |st(v)|$ is a homotopy equivalence because $st(v)$ is a cone, $|K \smallsetminus v| \hookrightarrow |K|$ is a homotopy equivalence by Theorem A.2.5.   □

Proposition A.2.7. *If Y is a contractible subcomplex of a CW-complex X, the quotient map $X \to X/Y$ is a homotopy equivalence.*

Proof. Since $Y \to *$ is a homotopy equivalence, by the gluing theorem so is $X \to X/Y$.   □

References

1. P.S. Alexandroff. *Diskrete Räume*. MathematiceskiiSbornik (N.S.) 2 (1937), 501–518.
2. J.J. Andrews, M.L. Curtis. *Free groups and handlebodies*. Proc. Amer. Math Soc. 16 (1965), 192–195.
3. M. Aschbacher and S. D. Smith. *On Quillen's conjecture for the p-groups complex*. Ann. of Math. 137 (1993), 473–529.
4. K. Baclawski and A. Björner. *Fixed Points in Partially Ordered Sets*. Adv. Math. 31 (1979), 263–287.
5. J.A. Barmak. *Algebraic topology of finite topological spaces and applications*. PhD Thesis, Facultad de Ciencias Exactas y Naturales, Universidad de Buenos Aires (2009).
6. J.A. Barmak. *On Quillen's Theorem A for posets*. J. Combin. Theory, Ser. A, 118 (2011), 2445–2453.
7. J.A. Barmak and E.G. Minian. *Minimal finite models*. J. Homotopy Relat. Struct. 2 (2007), No. 1, 127–140.
8. J.A. Barmak and E.G. Minian. *Simple homotopy types and finite spaces*. Adv. Math. 218 (2008), Issue 1, 87–104.
9. J.A. Barmak and E.G. Minian. *One-point reductions of finite spaces, h-regular CW-complexes and collapsibility*. Algebr. Geom. Topol. 8 (2008), 1763–1780.
10. J.A. Barmak and E.G. Minian. *Automorphism groups of finite posets*. Discrete Math., Vol. 309 (2009), Issue 10, 3424–3426.
11. J.A. Barmak and E.G. Minian. *Strong homotopy types, nerves and collapses*. Discrete Comput. Geom., doi:10.1007/s00454-011-9357-5, in press.
12. G. Barthel, J.P. Brasselet, K.H. Fieseler and L. Kaup. *Equivariant intersection cohomology of toric varieties*. Contemp. Math. 241 (1999), 45–68.
13. G. Birkhoff. *On groups of automorphisms*. Rev. Un. Mat. Argentina 11 (1946), 155–157.
14. A. Björner. *Posets, regular CW-complexes and Bruhat order*. Europ. J. Combinatorics 5 (1984), 7–16.
15. A. Björner. *Homotopy type of posets and lattice complementation*. J. Combin. Theory, Ser. A, 30 (1981), 90–100.
16. A. Björner. *Topological methods*. Handbook of Combinatorics, vol. 2 (1995), 1819–1872.
17. A. Björner. *Nerves, fibers and homotopy groups*. J. Combin. Theory, Ser. A, 102 (2003), 88–93.

18. A. Björner, M. Wachs and V. Welker. *Poset fiber theorems.* Trans. Amer. Math. Soc. 357 (2004), 1877–1899.
19. R. Boulet, E. Fieux and B. Jouve. *Simplicial simple-homotopy of flag complexes in terms of graphs.* European J. Combin. 31 (2010), 161–176.
20. K. Brown. *Euler characteristics of groups: The p-fractional part.* Invent. Math. 29 (1975), 1–5.
21. R. Brown. *Elements of modern topology.* McGraw-Hill, London, 1968.
22. E. Clader. *Inverse limits of finite topological spaces.* Homology, Homotopy Appl. 11 (2009), 223–227.
23. M.M. Cohen. *A Course in Simple Homotopy Theory.* Springer, New York, Heidelberg, Berlin (1970).
24. M.M. Cohen. *Whitehead torsion, group extensions and Zeeman's conjecture in high dimensions.* Topology 16 (1977), 79–88.
25. C.H. Dowker. *Homology groups of relations.* Ann. of Math. 56 (1952), 84–95.
26. D. Duffus, W. Poguntke and I. Rival. *Retracts and the fixed point problem for finite partially ordered sets.* Canad. Math. Bull 23 (1980), 231–236.
27. R. H. Fox. *On topologies for function spaces.* Bull. Amer. Math. Soc. 51 (1945), 429–432.
28. R. Fritsch and R.A. Piccinini. *Cellular Structures in Topology.* Cambridge Univ. Press, Cambridge (1990).
29. I. Gessel, R. Stanley. *Algebraic enumeration.* Handbook of Combinatorics, vol. 2 (1995), 1021–1061.
30. D. Gillman, D. Rolfsen. *The Zeeman conjecture for standard spines is equivalent to the Poincaré conjecture.* Topology 22 (1983), 315–323.
31. L. Glaser. *Geometrical combinatorial topology I.* Van Nostrand Reinhold, NY (1970).
32. B. Grünbaum. *Nerves of simplicial complexes.* Aequationes Math. 4 (1970), 63–73.
33. M. Hachimori. *Constructibility of Constructible Complexes* PhD Thesis, Graduate School of Arts and Sciences, the University of Tokyo (2000).
34. K.A. Hardie, S. Salbany, J.J.C. Vermeulen and P.J. Witbooi. *A non-Hausdorff quaternion multiplication.* Theoret. Comput. Sci. 305 (2003), 135–158.
35. K.A. Hardie and J.J.C. Vermeulen. *Homotopy theory of finite and locally finite T_0-spaces.* Exposition Math. 11 (1993), 331–341.
36. K.A. Hardie, J.J.C. Vermeulen and P.J. Witbooi. *A nontrivial pairing of finite T_0-spaces.* Topology Appl. 125 (2002), 533–542.
37. K.A. Hardie and P.J. Witbooi. *Crown multiplications and a higher order Hopf construction.* Topology Appl. 154 (2007), 2073–2080.
38. A. Hatcher. *Algebraic Topology.* Cambridge University Press (2002).
39. H. Honkasalo and E. Laitinen. *Equivariant Lefschetz classes in Alexander-Spanier cohomology.* Osaka J. Math. 33 (1996), 793–804.
40. G. Janelidze and M. Sobral. *Finite preorders and topological descent I.* J. Pure Appl. Algebra 175 (2002), 187–205.
41. J. Kahn, M. Saks and D. Sturtevant. *A topological approach to evasiveness.* Combinatorica 4 (1984), 297–306.
42. S. Kinoshita. *On some contractible continua without fixed point property.* Fund. Math. 40 (1953), 96–98.
43. S. Kono and F. Ushitaki. *Geometry of finite topological spaces and equivariant finite topological spaces.* In: Current Trends in Transformation Groups, ed. A. Bak, M. Morimoto and F. Ushitaki, 53–63, Kluwer Academic Publishers, Dordrecht (2002).
44. W.B.R. Lickorish. *Simplicial moves on complexes and manifolds.* Proceedings of the Kirbyfest (Berkeley, CA, 1998), 299–320. Geom. Topol. Monogr. 2, Geom. Topol. Publ., Coventry (1999).

45. A.T. Lundell and S. Weingram. *The topology of CW complexes*. Van Nostrand Reinhold Co. (1969).
46. F.H. Lutz. *Some results related to the evasiveness conjecture*. J. Combin. Theory Ser. B 81 (2001), no. 1, 110–124.
47. F.H. Lutz. *Examples of $\mathbb{Z}$-acyclic and contractible vertex-homogeneous simplicial complexes*. Discrete Comput. Geom. 27 (2002), no. 1, 137–154.
48. J. Matoušek. *LC reductions yield isomorphic simplicial complexes*. Contributions to Discrete Mathematics 3, 2 (2008), 37–39.
49. J. Matoušek, M. Tancer and U. Wagner. *Hardness of embedding simplicial complexes in $\mathbb{R}^d$*. J. Eur. Math. Soc. 13 (2011), 259–295.
50. S. Matveev. *Algorithmic topology and classification of 3-manifolds*. Second edition, Springer (2007).
51. J.P. May. *Finite topological spaces*. Notes for REU (2003). Available at http://www.math.uchicago.edu/~may/MISCMaster.html
52. J.P. May. *Finite spaces and simplicial complexes*. Notes for REU (2003). Available at http://www.math.uchicago.edu/~may/MISCMaster.html
53. J.P. May. *Finite groups and finite spaces*. Notes for REU (2003). Available at http://www.math.uchicago.edu/~may/MISCMaster.html
54. M.C. McCord. *Singular homology and homotopy groups of finite spaces*. Notices Amer. Math. Soc., vol. 12 (1965), p. 622.
55. M.C. McCord. *Singular homology groups and homotopy groups of finite topological spaces*. Duke Math. J. 33 (1966), 465–474.
56. M.C. McCord. *Homotopy type comparison of a space with complexes associated with its open covers*. Proc. Amer. Math. Soc. 18 (1967), 705–708.
57. J. Milnor. *Construction of universal bundles, II*. Ann. of Math. 63 (1956), 430–436.
58. J. Milnor. *Whitehead Torsion*. Bull. Amer. Math. Soc. 72 (1966), 358–426.
59. E.G. Minian. *Teorema de punto fijo de Lefschetz: Versión topológica y versiones combinatorias*. Unpublished notes. Available at http://mate.dm.uba.ar/~gminian/lefschetzfinal.pdf
60. E.G. Minian. *Some remarks on Morse theory for posets, homological Morse theory and finite manifolds*. arXiv:1007.1930
61. J.R. Munkres. *Elements of Algebraic topology*. Addison-Wesley (1984).
62. J.R. Munkres. *Topology, Second Edition*. Prentice Hall (2000).
63. R. Oliver. *Smooth fixed point free actions of compact Lie groups on disks*. Thesis, Princeton University (1974).
64. R. Oliver. *Fixed-Point Sets of Group Actions on Finite Acyclic Complexes*. Comm. Math. Helvetici 50 (1975), 155–177.
65. T. Osaki. *Reduction of finite topological spaces*. Interdiscip. Inform. Sci. 5 (1999), 149–155.
66. G. Perelman. *The entropy formula for the Ricci flow and its geometric applications*. arXiv:math/0211159v1
67. G. Perelman. *Ricci flow with surgery on three-manifolds*. arXiv:math/0303109v1
68. G. Perelman. *Finite extinction time for the solutions to the Ricci flow on certain three-manifolds*. arXiv:math/0307245v1
69. D. Quillen. *Higher algebraic K-theory, I: Higher K-theories*. Lect. Notes in Math. 341 (1972), 85–147.
70. D. Quillen. *Homotopy properties of the poset of nontrivial p-subgroups of a group*. Adv. Math. 28 (1978), 101–128.
71. I. Rival. *A fixed point theorem for finite partially ordered sets*. J. Combin. Theory A 21 (1976), 309–318.
72. D. Rolfsen. The *Poincaré conjecture and its cousins*. Unpublished notes. Available at http://www.math.ubc.ca/~rolfsen/papers/pccousins/PCcousins.pdf

73. G. Raptis. *Homotopy theory of posets*. Homology Homotopy Appl. 12 (2010), 211–230.
74. L.C. Siebenmann. *Infinite simple homotopy types*. Indag. Math. 32 (1970), 479–495.
75. E. Spanier. *Algebraic Topology*. Springer (1966).
76. R.E. Stong. *Finite topological spaces*. Trans. Amer. Math. Soc. 123 (1966), 325–340.
77. R.E. Stong. *Group actions on finite spaces*. Discrete Math. 49 (1984), 95–100.
78. M.C. Thornton. *Spaces with given homeomorphism groups*. Proc. Amer. Math. Soc. 33 (1972), 127–131.
79. M. Wachs. *Poset topology: Tools and applications*. In: E. Miller, V. Reiner and B. Sturmfels, Editors, Geometric Combinatorics, IAS/Park City Math. Ser. vol. 13, Amer. Math. Soc., Providence, RI (2007), 497–615.
80. J.W. Walker. *Homotopy type and Euler characteristic of partially ordered sets*. Europ. J. Combinatorics 2 (1981), 373–384.
81. J.W. Walker. *Topology and combinatorics of ordered sets*. Thesis, M.I.T., Cambridge, MA (1981).
82. C.T.C. Wall. *Formal deformations*. Proc. London Math. Soc. 16 (1966), 342–352.
83. V. Welker. *Constructions preserving evasiveness and collapsibility*. Discrete Math. 207 (1999), 243–255.
84. J.H.C. Whitehead. *Simplicial spaces, nuclei and m-groups*. Proc. London Math. Soc. 45 (1939), 243–327.
85. J.H.C. Whitehead. *On incidence matrices, nuclei and homotopy types*. Ann. of Math. 42 (1941), 1197–1239.
86. J.H.C. Whitehead. *Simple homotopy types*. Amer. J. Math. 72 (1950), 1–57.
87. E.C. Zeeman. *On the dunce hat*. Topology 2 (1964), 341–358.

List of Symbols

J.A. Barmak, *Algebraic Topology of Finite Topological Spaces and Applications*, Lecture Notes in Mathematics 2032, DOI 10.1007/978-3-642-22003-6, © Springer-Verlag Berlin Heidelberg 2011

Index

J.A. Barmak, *Algebraic Topology of Finite Topological Spaces and
Applications*, Lecture Notes in Mathematics 2032,
DOI 10.1007/978-3-642-22003-6, © Springer-Verlag Berlin Heidelberg 2011

LECTURE NOTES IN MATHEMATICS  Springer

Edited by J.-M. Morel, B. Teissier; P.K. Maini

Editorial Policy (for the publication of monographs)

1. Lecture Notes aim to report new developments in all areas of mathematics and their applications - quickly, informally and at a high level. Mathematical texts analysing new developments in modelling and numerical simulation are welcome.

   Monograph manuscripts should be reasonably self-contained and rounded off. Thus they may, and often will, present not only results of the author but also related work by other people. They may be based on specialised lecture courses. Furthermore, the manuscripts should provide sufficient motivation, examples and applications. This clearly distinguishes Lecture Notes from journal articles or technical reports which normally are very concise. Articles intended for a journal but too long to be accepted by most journals, usually do not have this "lecture notes" character. For similar reasons it is unusual for doctoral theses to be accepted for the Lecture Notes series, though habilitation theses may be appropriate.

2. Manuscripts should be submitted either online at www.editorialmanager.com/lnm to Springer's mathematics editorial in Heidelberg, or to one of the series editors. In general, manuscripts will be sent out to 2 external referees for evaluation. If a decision cannot yet be reached on the basis of the first 2 reports, further referees may be contacted: The author will be informed of this. A final decision to publish can be made only on the basis of the complete manuscript, however a refereeing process leading to a preliminary decision can be based on a pre-final or incomplete manuscript. The strict minimum amount of material that will be considered should include a detailed outline describing the planned contents of each chapter, a bibliography and several sample chapters.

   Authors should be aware that incomplete or insufficiently close to final manuscripts almost always result in longer refereeing times and nevertheless unclear referees' recommendations, making further refereeing of a final draft necessary.

   Authors should also be aware that parallel submission of their manuscript to another publisher while under consideration for LNM will in general lead to immediate rejection.

3. Manuscripts should in general be submitted in English. Final manuscripts should contain at least 100 pages of mathematical text and should always include

   - a table of contents;
   - an informative introduction, with adequate motivation and perhaps some historical remarks: it should be accessible to a reader not intimately familiar with the topic treated;
   - a subject index: as a rule this is genuinely helpful for the reader.

   For evaluation purposes, manuscripts may be submitted in print or electronic form (print form is still preferred by most referees), in the latter case preferably as pdf- or zipped psfiles. Lecture Notes volumes are, as a rule, printed digitally from the authors' files. To ensure best results, authors are asked to use the LaTeX2e style files available from Springer's web-server at:

   ftp://ftp.springer.de/pub/tex/latex/svmonot1/ (for monographs) and
   ftp://ftp.springer.de/pub/tex/latex/svmultt1/ (for summer schools/tutorials).

Additional technical instructions, if necessary, are available on request from lnm@springer.com.

4. Careful preparation of the manuscripts will help keep production time short besides ensuring satisfactory appearance of the finished book in print and online. After acceptance of the manuscript authors will be asked to prepare the final LaTeX source files and also the corresponding dvi-, pdf- or zipped ps-file. The LaTeX source files are essential for producing the full-text online version of the book (see http://www.springerlink. com/openurl.asp?genre=journal&issn=0075-8434 for the existing online volumes of LNM). The actual production of a Lecture Notes volume takes approximately 12 weeks.

5. Authors receive a total of 50 free copies of their volume, but no royalties. They are entitled to a discount of 33.3 % on the price of Springer books purchased for their personal use, if ordering directly from Springer.

6. Commitment to publish is made by letter of intent rather than by signing a formal contract. Springer-Verlag secures the copyright for each volume. Authors are free to reuse material contained in their LNM volumes in later publications: a brief written (or e-mail) request for formal permission is sufficient.

Addresses:
Professor J.-M. Morel, CMLA,
École Normale Supérieure de Cachan,
61 Avenue du Président Wilson, 94235 Cachan Cedex, France
E-mail: morel@cmla.ens-cachan.fr

Professor B. Teissier, Institut Mathématique de Jussieu,
UMR 7586 du CNRS, Équipe "Géométrie et Dynamique",
175 rue du Chevaleret
75013 Paris, France
E-mail: teissier@math.jussieu.fr

For the "Mathematical Biosciences Subseries" of LNM:

Professor P. K. Maini, Center for Mathematical Biology,
Mathematical Institute, 24-29 St Giles,
Oxford OX1 3LP, UK
E-mail : maini@maths.ox.ac.uk

Springer, Mathematics Editorial, Tiergartenstr. 17,
69121 Heidelberg, Germany,
Tel.: +49 (6221) 487-8259

Fax: +49 (6221) 4876-8259
E-mail: lnm@springer.com